The Coming Famine

The publisher gratefully acknowledges the generous support of the Barbara S. Isgur Public Affairs Endowment Fund of the University of California Press Foundation

THE COMING FAMINE

THE GLOBAL FOOD CRISIS AND WHAT WE CAN DO TO AVOID IT

Julian Cribb

UNIVERSITY OF CALIFORNIA PRESS
BERKELEY LOS ANGELES LONDON

University of California Press, one of the most distinguished university presses in the United States, enriches lives around the world by advancing scholarship in the humanities, social sciences, and natural sciences. Its activities are supported by the UC Press Foundation and by philanthropic contributions from individuals and institutions. For more information, visit www.ucpress.edu.

University of California Press
Berkeley and Los Angeles, California

University of California Press, Ltd.
London, England

Library of Congress Cataloging-in-Publication Data

Cribb, Julian.
The coming famine : the global food crisis and what we can do to avoid it / Julian Cribb.
p. cm.
Includes bibliographical references and index.
ISBN 978-0-520-27123-4 (pbk. : alk. paper)
1. Food supply. 2. Food industry and trade.
3. Sustainable agriculture. 4. Climatic changes.
I. Title.
HD9000.5.C715 2010
363.8—dc22

2010007683

Manufactured in the United States of America

19 18 17 16 15 14 13 12
10 9 8 7 6 5 4 3 2

This book is printed on Cascades Enviro 100, a 100% post consumer waste, recycled, de-inked fiber. FSC recycled certified and processed chlorine free. It is acid free, Ecologo certified, and manufactured by BioGas energy.

To the memory of Derek Tribe and Norman Borlaug, who believed that it is possible both to feed and to green the world

Contents

Illustrations

MAPS

FIGURES

TABLES

Preface

This book is a wake-up call.

It deals with the most urgent issue facing humanity in the twenty-first century, perhaps in all of history: the planetary emergency over whether or not we can sustain our food supply through the midcentury peak in human numbers, demand, and needs. It reflects on the likely consequences of our failure to do so.

Many, if not all, of the matters raised here have been voiced by various experts in agriculture, water, energy, trade, climate change, and other issues. Unfortunately, their voices have not been heeded as they should have been. The world has ignored the ominous constellation of factors that now make feeding humanity sustainably our most pressing task—even in times of economic and climatic crisis.

This book is written for the ordinary citizen of Planet Earth who wants to understand what our children and grandchildren face in their lifetimes—and what we must all do to avoid it.

It is written for all those who eat, and for those who intend that their children should eat in the future.

It doesn't claim to have all the answers: this is a remarkably complex and perturbing challenge. It does, however, gather together positive and practical ideas that show the way. And it reminds us that we humans are highly adaptive when our survival is at stake.

I am particularly grateful to the Crawford Fund for International

Agricultural Research and to Land and Water Australia, whose generous support enabled the book to be researched and written.

My warmest personal thanks are due to John Radcliffe, Richard Davies, John Williams, Meryl Williams, Des McGarry, Denis Blight, Cathy Reade, Bruce Wright, Jackie Hughes, Kadambot Siddique, Raj Patel, Iin Luther, Thomas Lumpkin, Nathan Russell, Alan Dupont, Andy Stoeckel, Arthur Blewett, James Ingram, Richard Meylan, and Graeme Pearman.

My especial appreciation goes to Reed Malcolm, Michael Renton, Elizabeth Berg, Madeleine Adams, Polly Kummel, and the University of California Press, who saw the need clearly and stuck with it courageously.

Julian Cribb
Canberra, 2010

ONE

WHAT FOOD CRISIS?

Lo que separa la civilización de la anarquía son solo siete comidas. (Civilization and anarchy are only seven meals apart.)

—Spanish proverb

Digging into a mountain of caviar, sea urchin roe, succulent Kyoto beef, rare conger eels, truffles, and fine champagne, the leaders of the world's richest and most powerful countries shook their heads over soaring grocery prices in the developed world and spreading hunger in Africa, India, and Asia. Over an eighteen-course banquet prepared for them by sixty chefs, the eight global potentates declared, "We are deeply concerned that the steep rise in global food prices coupled with availability problems in a number of developing countries is threatening global food security. The negative impacts of this recent trend could push millions more back into poverty."[1]

This statement, which followed the July 2008 meeting of the G8 (Group of Eight) nations in Hokkaido, Japan, was revelatory in several ways. The leaders of France, the United States, Russia, Britain, Germany, Canada, Italy, and Japan seemed bemused by the sudden emergence of the specter of food scarcity after decades of apparent abundance and cheap prices. This was a problem they clearly thought had been fixed.

Concealed within their response were embarrassing admissions. First, in urging major increases in global food aid, the leaders appeared to tacitly concede that wealthy countries had failed to fulfill their pledges to the United Nations's Millennium Development Goals of 2000 to fight poverty. Second, in calling on the world to reverse declining support for agricultural development and research, they were implicitly confessing that they had let these deteriorate. Third, in demanding food security early warning systems, the G8 leaders effectively admitted that they had been caught unawares by the emerging food crisis—and didn't like it. There are few things a politician likes less than an unforeseen development, so for good measure they backhanded the United Nations Food and Agriculture Organization (FAO), demanding its "thorough reform," presumably for the sin of having failed to get their attention with its previous warnings.[2]

The "Blessings of the Earth and the Sea Social Dinner" for the G8 leaders, hosted by the government of Japan, had more than a touch of the fall of the Roman Empire about it. The Earth's eight most powerful leaders and their partners regaled themselves on cornbread stuffed with caviar, smoked salmon, and sea urchin roe; hot onion tart and winter lily bulbs followed by kelp-flavored cold Kyoto beef with asparagus dressed with sesame cream; diced fatty tuna flesh with avocado, shiso, and jellied soy sauce; boiled clam, tomato, and shiso in jellied clear soup; water shield and pink conger dressed with a vinegary soy sauce; boiled prawn with jellied tosazu vinegar; grilled eel rolled in burdock; sweet potato; and fried and seasoned goby with soy sauce and sugar. This beginning was followed by a bisque of hairy crab and salt-grilled bighand thornyhead with vinegar-pepper sauce. The main course was poele of milk-fed lamb flavored with aromatic herbs and mustard, as well as roasted lamb with black truffle and pine seed oil sauce. This was followed by a special cheese selection with lavender honey and caramelized nuts, and then a whimsical "G8 fantasy dessert" and coffee with candied fruits and vegetables. The food was accompanied by Le Rêve grand cru/La Seule Gloire champagne; a sake wine, Isojiman Junmai Daiginjo Nakadori; Corton-Charlemagne 2005 (France); Ridge California Monte Bello 1997; and Tokaji Esszencia 1999 (Hungary).[3] The cost of holding the G8 summit (five hundred million dollars) could have fed for a week the additional one hundred million people left hungry by the emerging food crisis.[4]

With eloquent symbolism, this Petronian banquet made clear that the well-off part of humanity has largely forgotten what it is to go hungry and is awakening to an unpleasant shock: starvation and the wars, refugee crises, and collapse of nation-states that often accompany hunger have not been permanently banished after all. Indeed, they are once more at our doorstep. Food insecurity and its deadly consequences are again a pressing concern for every nation and each individual.

Despite the global food crisis of 2007–8, the coming famine hasn't happened yet. It is a looming planetary emergency whose interlocked causes and deeper ramifications the world has barely begun to absorb, let alone come to grips with. Experts predict that the crisis will peak by the middle of the twenty-first century; it is arriving even faster than climate change. Yet there is still time to forestall catastrophe.

The first foreshocks were discernible soon after the turn of the millennium. In the years from 2001 to 2008 the world steadily consumed more grain than it produced, triggering rising prices, growing shortages, and even rationing and famine in poorer countries. The global stockpile of grain shrank from more than a hundred days' supply of food to less than fifty days'.[5] It was the difference between a comfortable surplus and alarming shortages in some countries; it was accompanied by soaring prices—and the resulting fury of ordinary citizens.

It was mainly this simple fact of each year consuming slightly more than we grew that panicked the long-quiescent grain markets, triggering a cycle of price increases that sent shockwaves through consumers in all countries, governments, and global institutions such as the United Nations, its FAO, and the World Bank. All of a sudden, food security, having been off the political menu for decades, was heading the bill of fare—not even to be entirely eclipsed by the spectacular crash of the world's financial markets that followed soon afterward.

That the world was suddenly short of food—after almost a half century of abundance, extravagant variety, year-round availability, and the cheapest real food prices enjoyed by many consumers in the whole of human history—seemed unimaginable. On television, celebrity chefs extolled the virtue of devouring animals and plants increasingly rare in the wild; magazines larded their pages with mouth-watering recipes to tempt their overfed readers' jaded appetites; food corporations churned out novel concoctions of salt, sugar, fat, emulsifier, extender, and dye; fast-food outlets disgorged floods of dubious nutrition to fatten an already overweight 1.4 billion people. And, in the third world, nearly fifteen thousand children continued to die quietly and painfully each day from hunger-related disease.[6]

"A brutal convergence of events has hit an unprepared global market, and grain prices are sky high. The world's poor suffer most," stated the *Washington Post*. "The food price shock now roiling world markets is destabilizing governments, igniting street riots and threatening to send a new wave of hunger rippling through the world's poorest nations. It is outpacing even the Soviet grain emergency of 1972–75, when world food prices rose 78 percent." Between 2005 and 2008, food prices rose on average by 80 percent, according to the FAO.[7]

"Rocketing food prices—some of which have more than doubled in two years—have sparked riots in numerous countries recently," *Time* magazine reported. "Millions are reeling . . . and governments are scrambling to staunch a fast-moving crisis before it spins out of control. From Mexico to Pakistan, protests have turned violent." *Time* attributed events to booming demand from newly affluent Chinese and Indian consumers, freak weather events that had reduced harvests, the spike in oil prices, and growth in the production of farm biofuels.[8]

In early 2007, thousands of Mexicans turned out on the streets in protest over the "tortilla crisis"—savage increases in the cost of maize flour. Over the ensuing months food riots or public unrest over food prices were reported by media in Haiti, Malaysia, Indonesia, the Philippines, Bangladesh, India, Burkina Faso, Senegal, Cameroon, Morocco, Mauritania, Somalia, Ethiopia, Madagascar, Kenya, Egypt, Ivory Coast, Yemen, the United Arab Emirates, Mexico, and Zimbabwe. In Haiti riots forced the resignation of the prime minister and obliged the United Nations World Food Programme to provide emergency aid to 2.3 million people. The new government of Nepal tottered. Mexico announced plans to freeze the prices of 150 staple foods. The U.K. *Guardian* reported riots in fifteen countries; the *New York Times* and the World Bank both said thirty. The FAO declared that thirty-seven countries faced food crises due to conflict or disaster at the start of 2008, adding that 1.5 billion people living in degraded lands were at risk of starvation. The *Economist* magazine succinctly labeled it a "silent tsunami."[9]

The rhetoric reflected the sudden, adventitious nature of the crisis. "It is an apocalyptic warning," pronounced Tim Costello, the Australian head of the aid agency World Vision. "Until recently we had plenty of food: the question was distribution. The truth is because of rising oil prices, global warming and the loss of arable land, all countries that can produce food now desperately need to produce more."[10]

"What we are witnessing is not a natural disaster—a silent tsunami or a perfect storm. It is a man-made catastrophe," the World Bank group president Robert Zoellick advised the G8 leaders feasting in Japan. Major rice-growing countries, including India, Vietnam, China, and Cambodia, imposed export restrictions to curb rice price inflation at home. Malaysia, Singapore, Sri Lanka, and the Philippines began stockpiling grain while Pakistan and Russia raised wheat export taxes and Brazil,

Indonesia, and Argentina imposed export restrictions. Guinea banned all food exports.[11]

The panic reached a peak in Asia, where rice prices soared by almost 150 percent in barely a year. "Nobody has ever seen such a jump in the price of rice," said sixty-eight-year-old Kwanchai Gomez, the executive director of the Thai Rice Foundation. Filipino fast-food outlets voluntarily reduced customer portions by half. In Thailand, thieves secretly stripped rice paddies by night to make a fast profit. India banned the export of all non-basmati rice, and Vietnam embargoed rice exports, period, sending Thai rice prices spiraling upward by 30 percent. The giant U.S. retailer Wal-Mart rationed rice sales to customers of its Sam's Club chain; some British retailers did likewise. Such measures did little to quell the panic, which was originally touched off by a 50 percent drop in surplus rice stocks over the previous seven years. The International Rice Research Institute attributed the crisis to loss of land to industrialization and city sprawl, the growing demand for meat in China and India, and floods or bad weather in Indonesia, Bangladesh, Vietnam, China, and Burma.[12]

By mid-2009, accelerated by the worldwide financial crash, thirty-three countries around the world were facing either "alarming" or "extremely alarming" food shortages, a billion people were eating less each day[13]—and most of Earth's citizens were feeling the pinch. Though food prices fell, alongside prices of stocks and most other commodities, in the subsequent months, they fell only a little—and then began to rise again.

What happened in 2008 wasn't the coming famine of the twenty-first century, merely a premonition of what lies ahead. This will not be a single event, affecting all nations and peoples equally at all times, but in one way or another it will leave no person in the world untouched. The reemergence of food scarcity occurs after decades of plenty, accompanied by the lowest real food prices for consumers in history. These bounteous years were the consequence of a food production miracle achieved by the world's farmers and agricultural scientists from the 1960s on—a miracle of which the urbanized world of today seems largely oblivious and which we have forgotten to renew.

By the early twenty-first century, signs of complacency were in evidence. In 2003, a conference of the Consultative Group on International Agricultural Research in Nairobi was told, "According to the Food and Agriculture Organization of the United Nations, the number of food-insecure people in developing countries fell from 920 million in 1980 to

799 million in 1999." Even in the immediate aftermath of the 2008 food price spike, the FAO itself, along with the Organization for Economic Cooperation and Development, remarked, "the underlying forces that drive agricultural product supply (by and large productivity gains) will eventually outweigh the forces that determine stronger demand, both for food and feed as well as for industrial demand, most notably for biofuel production. Consequently, prices will resume their decline in real terms, though possibly not by quite as much as in the past."[14]

For some years, reassuring statements such as these had been repeatedly aired in the food policy, overseas aid, and research worlds. Unintentionally, food scientists and policy makers were sending a signal to governments and aid donors around the world that implied, "Relax. It's under control. We've fixed the problem. Food is no longer critical." Not surprisingly, aid donors rechanneled scarce funds to other urgent priorities—and growth in crop yields sagged as the world's foot came off the scientific accelerator.

Many found the new crisis all the more mysterious for its apparent lack of an obvious trigger. Various culprits were pilloried by blame-seeking politicians and media. Biofuels, after being talked up as one of the great hopes for combating climate change, quickly became a villain accused of "burning the food of the poor," and from China to Britain, countries slammed the brakes on policies intended to encourage farmers to grow more "green fuel" from grain. According to the World Bank, biofuels could have caused as much as three-quarters of the hike in food prices. Equally to blame, according to other commentators, were oil prices, which had soared sixfold in the five years from mid-2003 to mid-2008 (although they fell again sharply as the global recession bit deep), with severe consequences for the cost of producing food, through their impact on farmers' fuel, fertilizer, pesticide, and transportation costs. In developed countries the financial pain was high, but in developing nations it was agony: farmers simply could not afford to buy fertilizer, and crop yields began to slip. In Thailand rice farmers quietly parked their new but unaffordable tractors in their sheds and went back to plowing with buffalo; buffalo breeders experienced a bonanza. "Energy and agricultural prices have become increasingly intertwined," commented Joachim von Braun, the head of the International Food Policy Research Institute. "High energy prices have made agricultural production more expensive by raising the cost of cultivation, inputs—especially fertilizers and irrigation—and transportation of inputs and outputs. In poor countries, this hinders production response to high output prices. The main new link between energy and agricultural prices, however, is the competition of grain and oilseed land for feed and food, versus their use for bio energy."[15]

Speculators, fleeing crumbling financial markets and discovering an unlikely haven in booming agricultural commodities, were a favorite target of media ire: "Food was becoming the new gold. Investors fleeing Wall Street's mortgage-related strife plowed hundreds of millions of dollars into grain futures, driving prices up even more. By Christmas (2007), a global panic was building," reported the *Washington Post*. In developing nations, traders and grain dealers were accused of buying up surplus stocks and hoarding them to drive the prices higher still. In the Philippines the government threatened hoarders with charges of economic sabotage and sent armed soldiers to supervise the distribution of subsidized grain.[16] Retirement and hedge funds, casting about for something to invest in that wasn't going to hell in a handbasket, also jumped on farm commodities and even agribusiness enterprises—areas such investors traditionally shun.

Many saw the crisis as simply a result of the growth of human population, the inexorable climb from 3 billion people in 1960 to 6.8 billion by 2008—the hundred million more mouths we have to feed in each succeeding year. Others ascribed it chiefly to burgeoning appetites in China and India, which had in a matter of five years or so together added the consumer equivalent of Europe to global demand for food as their emergent middle classes indulged in the delights of diets containing far more meat, poultry, dairy, and fish than ever before. In China, meat consumption trebled in less than fifteen years, requiring a tenfold increase in the grain needed to feed the animals and fish. One way to visualize the issue is that growth in global food production of 1–1.5 percent a year has more or less kept pace with growth in population—but has fallen short of meeting the growth in demand. One explanation for this is that farmers around the world have not responded by increasing the area of land they plant and harvest or raising their crop yields so rapidly as in the past. The big question is: why?

Some blamed the weather. Portentously, many were quick to discern the looming shadow of climate change in the run of droughts, floods, and other natural mishaps that had disrupted global farm production across most continents in recent years. In eastern Australia a ten-year drought slashed grain production and all but obliterated the rice industry; the unprecedented draining of Australia's food bowl, the Murray-Darling Basin, threatened to eliminate fruit, vegetable, and livestock industries reliant on irrigation. Similar hardship faced producers across sub-Sahelian Africa. Floods in China and along the Mississippi River wreaked local havoc with grain production. In Burma, Cyclone Nargis flattened the Irrawaddy Delta rice crop, propelling Asian prices into a fresh spiral. Heat waves in California and torrential rains in India added to perceptions—heightened by media reportage—that the climate was running amok.[17]

Other commentators sought villains among the world's governments, blaming protectionism and hidden trade barriers, farm subsidies, food price controls or taxes, environmental and health restrictions, the ensnaring of farmers in snarls of red tape, along with the perennial failure of trade negotiators to open up global trade in agricultural products. Supermarkets and globalization of the food trade came in for flak, especially from the political left and from farmers themselves, for driving down farm commodity prices and thus discouraging growers from increasing production. Economic observers read the crisis as primarily due to weaker growth in food production at a time of strong growth in consumer demand, especially in China and India and among affluent populations worldwide.[18]

The Green Revolution, whose technologies had delivered the last great surge in global food production in the 1970s and 1980s, seemed to be fizzling out, a view supported by the disturbing slide in crop yield advances. Yields of the major crops of wheat, maize, and rice had once increased by as much as 5 and even 10 percent a year—now they were increasing by 1 percent or nothing at all. In the overheated economy of the early twenty-first century, farm costs had soared along with oil prices, hindering farmers from adopting newer, but costlier and more energy-intensive, technologies. In advanced countries, some scientists whispered, we might actually be approaching the physical limits of the ability of plants to turn sunlight into edible food.

In the general hunt for someone to blame for the short-term food crisis, a more profound truth was being obscured—that the challenge is far deeper, longer-term, and more intractable than most people, and certainly most governments, understand. It stems from the magnifying and interacting constraints on food production generated as civilization presses harder against the finite bounds of the planet's natural resources, combined with human appetites that seem to know no bounds.

This challenge is more pressing even than climate change. A climate crisis may emerge over decades. A food crisis can explode within weeks—and kill within days. But the two are also interlocked. "If the world were to experience a year of bad weather similar to that experienced in 1972, the current 'food crisis' would pale in comparison to the crisis that would arise as a result. This should be taken as a warning that advance planning ought to be done if total chaos is to be avoided," observes the resource analyst Bruce Sundquist.[19]

The character of human conflict has also changed: since the early 1990s, more wars have been triggered by disputes over food, land, and water than over mere political or ethnic differences. This should not surprise us: people have fought over the means of survival for most of history. But in the abbreviated reports on the nightly media, and even in the rarefied realms of government policy, the focus is almost invariably on the players—the warring national, ethnic, or religious factions—rather than on the play, the deeper subplots building the tensions that ignite conflict. Caught up in these are groups of ordinary, desperate people fearful that there is no longer sufficient food, land, and water to feed their children—and believing that they must fight "the others" to secure them. At the same time, the number of refugees in the world doubled, many of them escaping from conflicts and famines precipitated by food and resource shortages. Governments in troubled regions tottered and fell.

The coming famine is planetary because it involves both the immediate effects of hunger on directly affected populations in heavily populated regions of the world in the next forty years—and also the impacts of war, government failure, refugee crises, shortages, and food price spikes that will affect all human beings, no matter who they are or where they live. It is an emergency because unless it is solved, billions will experience great hardship, and not only in the poorer regions. Mike Murphy, one of the world's most progressive dairy farmers, with operations in Ireland, New Zealand, and North and South America, succinctly summed it all up: "Global warming gets all the publicity but the real imminent threat to the human race is starvation on a massive scale. Taking a 10–30 year view, I believe that food shortages, famine and huge social unrest are probably the greatest threat the human race has ever faced. I believe future food shortages are a far bigger world threat than global warming."[20]

The coming famine is also complex, because it is driven not by one or two, or even a half dozen, factors but rather by the confluence of many large and profoundly intractable causes that tend to amplify one another. This means that it cannot easily be remedied by "silver bullets" in the form of technology, subsidies, or single-country policy changes, because of the synergetic character of the things that power it.

To see where the answers may lie, we need to explore each of the main drivers. On the demand side the chief drivers are:

Population. Although the rate of growth in human numbers is slowing, the present upward trend of 1.5 percent (one hundred million more people) per year points to a population of around 9.2 billion in 2050—3 billion more than in 2000. Most of this expansion will take place in poorer countries and in tropical/subtropical regions. In countries where birth rates are falling, governments are bribing their citizens with subsidies to have more babies in an effort to address the age imbalance.

Consumer demand. The first thing people do as they climb out of poverty is to improve their diet. Demand for protein foods such as meat, milk, fish, and eggs from consumers with better incomes, mainly in India and China but also in Southeast Asia and Latin America, is rising rapidly. This in turn requires vastly more grain to feed the animals and fish. Overfed rich societies continue to gain weight. The average citizen of Planet Earth eats one-fifth more calories than he or she did in the 1960s—a "food footprint" growing larger by the day.

Population *and* demand. This combination of population growth with expansion in consumer demand indicates a global requirement for food by 2050 that will be around 70–100 percent larger than it is today.[21] Population and demand are together rising at about 2 percent a year, whereas food output is now increasing at only about 1 percent a year.

These demand-side factors could probably be satisfied by the world adopting tactics similar to those of the 1960s, when the Green Revolution in farming technology was launched, were it not for the many constraints on the supply side that are now emerging to hinder or prevent such a solution:

Water crisis. Put simply, civilization is running out of freshwater. Farmers presently use about 70 percent of the world's readily available freshwater to grow food. However, megacities, with their huge thirst for water for use in homes, industry, and waste disposal, are increasingly competing with farmers for this finite resource and, by 2050, these uses could swallow half or more of the world's available freshwater at a time when many rivers, lakes, and aquifers will be drying up.[22] Unless major new sources or savings are found, farmers will have about half of the world's currently available freshwater with which to grow twice the food.

Land scarcity. The world is running out of good farmland. A quarter of all land is now so degraded that it is scarcely capable of yielding food. At the same time, cities are sprawling, smothering the world's most fertile soil in concrete and asphalt, while their occupants fan out in search of cheap land for recreation that diverts the best food-producing areas from agriculture. A third category of land is poisoned by toxic industrial pollution. Much former urban food production has now ceased. The emerging global dearth of good farmland represents another severe limit on increasing food production.

Nutrient losses. Civilization is hemorrhaging nutrients—substances essential to all life. Annual losses in soil erosion alone probably exceed all the nutrients applied as fertilizer worldwide. The world's finite nutrient supplies may already have peaked. Half the fertilizer being used is wasted. In most societies, up to half the food produced is trashed or lost; so too are most of the nutrients in urban waste streams. The global nutrient cycle, which has sustained humanity throughout our history, has broken down.

Energy dilemma. Advanced farming depends entirely on fossil fuels, which are likely to become very scarce and costly within a generation. At present farmers have few alternative means of producing food other than to grow fuel on their farms—which will reduce food output by 10–20 percent. Many farmers respond to higher costs simply by using less fertilizer or fuel—and so cutting yields. Driven by high energy prices and concerns about climate change, the world is likely to burn around 400 million tonnes (441 million U.S. tons) of grain as biofuels by 2020[23]—the equivalent of the entire global rice harvest.

Oceans. Marine scientists have warned that ocean fish catches could collapse by the 2040s due to overexploitation of wild stocks.[24] Coral reefs—whose fish help feed about five hundred million people—face decimation under global warming. The world's oceans are slowly acidifying as carbon dioxide from the burning of fossil fuels dissolves out of the atmosphere, threatening ocean food chains. Fish farms are struggling with pollution and sediment runoff from the land. The inability of the fish sector to meet its share of a doubling in world food demand will throw a heavier burden onto land-based meat industries.

Technology. For three decades the main engine of the modern food miracle, the international scientific research that boosted crop yields, has been neglected, leading to a decline in productivity gains. Farmers worldwide are heading into a major technology pothole, with less new knowledge available in the medium run to help them to increase output.

Climate. The climate is changing: up to half the planet may face regular drought by the end of the century. "Unnatural disasters"—storms, floods, droughts, and sea-level rise—are predicted to become more frequent and intense, with adventitious impacts on food security, refugee waves, and conflict.

Economics, politics, and trade. Trade barriers and farm subsidies continue to distort world markets, sending the wrong price signals to farmers and discouraging investment in agriculture and its science. The globalization of food has helped drive down prices received by farmers. Speculators have destabilized commodity markets, making it riskier for farmers to make production decisions. Some countries discourage or ban food exports and others tax them, adding to food insecurity. Others pay their farmers to grow fuel instead of food. A sprawling web of health, labor, and environmental regulation is limiting farmers' freedom to farm.

The collapse in world economic conditions in late 2008 and 2009 has changed the prices of many things, including land, food, fuel, and fertilizer—but has not altered the fact that demand for food continues to grow while limits on its production multiply. Indeed, the economic crash exacerbated hunger among the world's poor, and has not altered the fundamentals of climate change, water scarcity, population growth, land degradation, or nutrient or oil depletion.

In early 2009 a report by Chatham House, a think tank focused on international affairs, observed that a lower food price "does not mean that policy-makers around the world can start to breathe a sigh of relief. . . . [E]ven at their somewhat diminished levels current prices remain acutely problematic for low-income import-dependent countries and for poor people all over the world. The World Bank estimates that higher food prices have increased the number of undernourished people by as much as 100 million from its pre-price-spike level of 850 million."[25]

In the medium and longer term, the report warned, food prices were poised to rise again. "Although many policy-makers have taken a degree of comfort from a recent OECD-FAO report on the world's agricultural outlook to 2017 . . . the report largely overlooked the potential impact of long-term resource scarcity trends, notably climate change, energy security and falling water availability."[26]

To sum it all up, the challenge facing the world's 1.8 billion women and men who grow our food is to double their output of food—using far less water, less land, less energy, and less fertilizer. They must accomplish this on low and uncertain returns, with less new technology available, amid more red tape, economic disincentives, and corrupted markets, and in the teeth of spreading drought. Achieving this will require something not far short of a miracle.

Yet humans have done it before and, resilient species that we are, we can do it again. This time, however, it won't just be a problem for farmers, scientists, and policy makers. It will be a challenge involving every single one of us, in our daily lives, our habits, and our influence at the ballot box and at the supermarket.

It will be the greatest test of our global humanity and our wisdom we have yet faced.

TWO

FOOD . . . OR WAR?

When the Cold War ended, we expected an era of peace. . . . What we got was a decade of war.

—President Jimmy Carter

Former U.S. president Jimmy Carter was among the first to notice that something big had shifted in the world's geopolitical gravity field when he penned these words in 1999.[1] With the instincts of a veteran statesman as well as those of a farmer, Carter perceived that hunger wasn't just a poverty issue—it was an emerging global security risk.

If large regions of the world run short of food, land, or water in the decades that lie ahead, then wholesale, bloody wars are liable to follow. These wars have already begun, although many of today's governments and media seem unconscious of the fact.

We should not be surprised. Famine and war have been inseparable Horsemen of the Apocalypse since antiquity. In the modern era famine notably propelled events as significant as the French Revolution, where what started as a bread crisis ultimately claimed a half million lives in the ensuing civil war and its civilian massacres; and the Russian Revolution, where food protests unleashed a civil war that devoured nine million human lives between 1917 and 1922.[2] Even World War II had an imponderable component in the struggle for productive land—or *lebensraum,* as Nazi philosophy defined it. Yet food, land, and water are nowadays widely disregarded as the wellsprings of war.

Carter continued:

> The devastation occurs primarily in countries whose economies depend on agriculture but lack the means to make their farmland productive. These are developing countries such as Sudan, Congo, Colombia, Liberia, Peru, Sierra Leone and Sri Lanka. . . .
>
> The economies of Europe, the United States, Canada and Japan were built on strong agriculture. But many developing countries have shifted their priorities away from farming in favor of urbanization, or they have reduced investments in agriculture because of budget shortages. At the same time, industrialized countries continue to cut their foreign aid budgets, which fund vital scientific research and extension work to improve farming in developing countries.

"The message is clear," he concluded. "There can be no peace until people have enough to eat. Hungry people are not peaceful people."[3]

For decades many academics and policy makers have assumed that war is the parent and famine its child, yet recent conflicts in which critical food shortages have played a part in igniting events have begun to beg the question, Is it war that drives famine, or do scarcities of food, land, and water also sometimes lead to war? Scholars have closely dissected the chicken, but few so far have probed the egg—yet this may be critical to an understanding of one of the primary forces shaping our times.[4]

The shift began almost imperceptibly in 1999, when a groundbreaking study by scholars affiliated with the International Peace Research Institute of Oslo, an independent think tank devoted to research on global conflict, concluded that with the ending of the Cold War, "the new internal wars, extremely bloody in terms of civilian casualties, reflect subsistence crises and are largely apolitical."[5] This hinted, for the first time, that resource scarcity of food, land, and water could become a major trigger for conflict rather than merely a consequence of it. (The prevailing expert view, however, still mainly considers scarcity a consequence.) At the dawn of the century of humanity's greatest resource scarcities it was a serious wake-up call, yet one through which many slumbered on.

"The crises stem from the failure of development, the loss of livelihood and the collapse of states. These factors add up to a vicious cycle," the Oslo scholars Indra de Soysa and Nils Petter Gleditsch explained. "The causes of armed conflict are perpetuated by conflict itself. People fight over vital necessities such as food, to protect a livelihood. . . . [S]tates that can provide such necessities also create conditions conducive to peace and prosperity."[6] Peace and prosperity, in turn, create the conditions necessary for democratic government, civil society, and a culture of peace, they added. Democracy is not commonly thought of as a food by-product, but it probably is.

In their study, de Soysa and Gleditsch published a disturbing map (see map 1). It showed all the countries of the world where food production was most critical to the survival of the nation-state—and all the places where, in the previous ten years, war and strife had broken out. The coincidence was more than striking. If your country is at the mercy of a shaky food supply, the map implies, watch out for war.

The opposite was equally evident: those places where food was plentiful—"old" Europe, North America, Australasia, and parts of Latin America—had escaped mass bloodletting within their own territories during the decade. Peace, the study implied, prefers a full platter.

The causes of these wars included disputes between new settlers and existing landholders, unjust land distribution due to corrupt ownership or government, environmental degradation so bad as to reduce the food supply, lack of access to water, and famine. Environmental wars, so far, are rare—but several commentators think that such conflicts may become more frequent as humanity presses against the limits of the Earth's resources. "Conditions affecting the livelihoods of the majority of people in poor countries are at the heart of the internal violence. The inability to meet food requirements drives people to adopt alternative survival strategies, one of which is to join rebellions and criminal insurgencies. In such situations the use of violence is primarily for economic goals, rather than the political ends that drove many revolutionary movements during the Cold War," de Soysa and Gleditsch wrote.[7]

Added to this may be another factor more primal still: love of one's children. Of all the indicators that point most reliably to government collapse and the probability of conflict, none is more brutally eloquent than the death rate among children. Starvation and malnutrition-related disease are the main causes of high infant mortality. Those countries with the most child deaths also have high levels of conflict.[8]

Love of children, horribly, is what may—at times—furnish the motivation for genocide: the blind desire to exterminate "the other," to eliminate the competition they pose for the basics of life. The roots of the 1994 Rwandan genocide were on the farm: "The country relied heavily on coffee exports for hard currency and government revenues. The collapse of world prices in the early 1990s led to high unemployment, reduced farm incomes, reduced social spending, and a citizenry receptive to government incitement of ethnic and political violence."[9] Most studies still focus on the salient features of genocide rather than its underlying

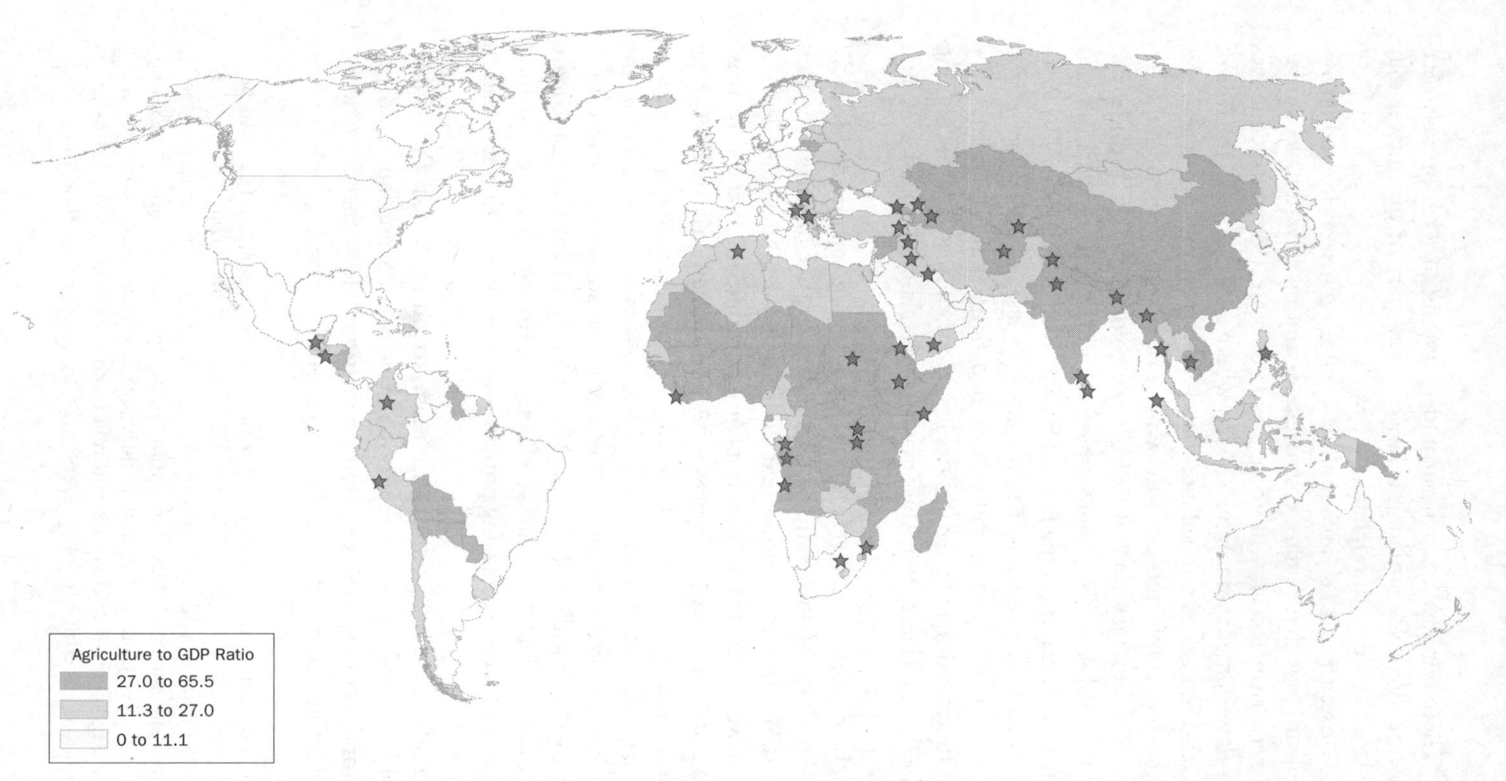

Map 1 Relationship between food insecurity and armed conflict. Where nations' food supplies are at risk (dark areas), there is a high chance of conflict (stars). Source: Indra de Soysa and Nils Petter Gleditsch with Michael Gibson, Margareta Sollenberg, and Arthur H. Westing, "To Cultivate Peace: Agriculture in a World of Conflict," International Peace Research Institute (PRIO) Report 1/99, Oslo, 1999, fig. 2, p. 102, www.isn.ethz.ch/isn/Digital-Library/Publications/Detail/?ots591=CAB359A3–9328–19CC-A1D2–8023E646B22C&lng=en&id=37962.

drivers. There remains a gap in our understanding of what propels societies toward this self-mutilating behavior—and part of the answer may well lie in scarcities of food, land, and water.

Some observers also claim a link between food insecurity and terrorism, pointing out that hungry countries are among those most likely to furnish terrorism recruits. In 2002, heads of state from fifty countries met at a development summit in Mexico where they discussed the role of poverty and hunger as a breeding ground for terrorism. "No-one in this world can feel comfortable or safe while so many are suffering and deprived," UN secretary general Kofi Annan told them. The president of the UN General Assembly, Han Seung-Soo, added that the world's poorest countries were a breeding ground for violence and despair. The Peruvian president Alejandro Toledo added, "To speak of development is to speak also of a strong and determined fight against terrorism."[10] Around the world many guerrilla and insurgent causes—such as Shining Path, the Tamil Tigers, and Abu Sayyaf—have claimed injustice in land ownership and use as one of their motivating causes.

A lack of water is a key factor in encouraging terrorism. Mona El Kody, the chair of the National Water Research Unit in Egypt told the Third World Water Forum that living without an adequate level of access to water created a "non-human environment" that led to frustration, and from there to terrorism. "A non-human environment is the worst experience people can live with, with no clean water, no sanitation," she said, adding that this problem was at its most acute in the Middle East, where 1 percent of the world's freshwater is shared by 5 percent of the world's population. Ms. El Kody added that inadequate water resources had the additive effect of reducing farming and food production, thereby increasing poverty—another factor that can lead to terrorism.[11]

Most of the "new" conflicts are to be found in Africa, the Middle East, and parts of Asia—the result of a cycle of constant famine, deprivation, and periodic violence, leading in inevitable sequence to worse hunger, greater deprivation, and more vicious fighting.

> Food and economic insecurity and natural resource scarcities . . . can be major sources of conflict. When politically dominant groups seize land and food resources, deny access to other culturally or economically marginalized groups, and cause hunger and scarcities, violence often flares. In Ethiopia, Rwanda, and Sudan, food crises resulting from drought and mismanagement of agriculture and relief and development aid led to rebellion and government collapse, followed by even greater food shortfalls

> in ensuing years of conflict. Denial of the right to food has been linked to uprisings and civil war in Central America and Mexico. Food insecurity is also integral to civil conflicts in Asia. Competition for resources has generated cycles of hunger and hopelessness that have bred violence in Sri Lanka as well as Rwanda.[12]

These afflicted regions are generally places disconnected from the global economic mainstream, where strong-man governments arise and just as quickly crumble, having only political quicksand on which to build a foundation for stability and progress. This is vital to an understanding of what is going wrong with global food production: in nearly all these countries, food is of the first importance, and only after you have enough food can you form a government stable enough to deliver water, health care, education, opportunity for women, justice, and economic development. By neglecting or reducing support for basic food production—as many have during the past twenty-five years—in order to spread aid across these equally deserving causes, the world's aid donors may unintentionally have laid the foundation for future government failure and conflict.

"The absolute number of countries with food crises caused by war and conflicts has increased since the 1980s as has the relative share of food crises caused by socio-economic factors from about 2 percent to 27 percent by 2007," the UN's Food and Agriculture Organization informed the Committee on World Food Security in 2008. "The recent sharp increase in the price of imported food commodities is an example of a socio-economic shock that can exacerbate or cause food crises in many countries."[13]

In describing the triggers for today's conflicts, the U.S. Central Intelligence Agency states in its online *World Fact Book,* "Territorial disputes may evolve from historical and/or cultural claims, or they may be brought on by resource competition. . . . [S]ources of contention include access to water and mineral (especially hydrocarbon) resources, fisheries, and arable land." It adds, "Armed conflict prevails not so much between the uniformed armed forces of independent states as between stateless armed entities that detract from the sustenance and welfare of local populations, leaving the community of nations to cope with resultant refugees, hunger, disease, impoverishment, and environmental degradation."[14]

The CIA presumably has a good grasp of what makes people fight. In its quest to predict where the world's next trouble spots are likely to erupt, it clearly recognizes—on the public record—the power of disputes over food, land, and water to fire the tinder. A scan of the scores of current disputes it lists soon reveals the frequent appearance of these three casus belli and of the border confrontations triggered by them. The question is why many of the world's governments appear unresponsive to the need to deal with these triggers for war, if these are the warnings their intelligence services are providing.

The threat of conflict over food, land, and water is not, however, confined to the marginal world. Increasingly it imperils the economic powerhouses of the global economy in the early twenty-first century. In 2001 the Australian strategic analyst Alan Dupont predicted, "Food is destined to have greater strategic weight and import in an era of environmental scarcity. While optimists maintain that the world is perfectly capable of meeting the anticipated increases in demand for essential foodstuffs, there are enough imponderables to suggest that prudent governments would not want to rely on such a felicitous outcome." Anticipating the food crisis of 2007–8 by several years, he presciently added, "East Asia's rising demand for food and diminishing capacity to feed itself adds an unpredictable new element to the global food equation for several reasons. The gap between production and consumption of key foodstuffs globally is narrowing dangerously and needs to be reversed." Bearing out his words, Singapore president Lee Hsieng Loong told a 2008 international defense conference, "In the longer term, the trends towards tighter supplies and higher prices will likely reassert themselves. This has serious security implications. The impact of a chronic food shortage will be felt especially by the poor countries. The stresses from hunger and famine can easily result in social upheaval and civil strife, exacerbating conditions that lead to failed states. Between countries, competition for food supplies and displacement of people across borders could deepen tensions and provoke conflict and wars."[15]

Indeed, the U.S. academics Ellen Messer and Marc Cohen argue that most modern conflicts ought to be viewed as "food wars," that is, "the practice of warring parties fighting for control of food supplies to reward their supporters and punish their enemies." In 2003 they estimated that there were fifty-six million people living in twenty-seven countries where food wars were taking place.[16]

Aid agencies understood this better than anyone else. When the 2007–8 food price crisis struck, the International Red Cross (IRC) immediately warned of the risk of a surge in violence. Jakob Kellenberger, the president of the IRC's International Committee, told media that "there is also the potential of food-related violence." Price hikes for staple foods had sparked riots in places such as Haiti, Egypt, and Somalia, and Kellenberger predicted that the neutral Red Cross would be facing even greater responsibility "when that violence reaches the level of an armed conflict. . . . You can imagine when you have countries where you have already an armed conflict, where you have already a high level of violence and you have at the same time a lot of poor and extremely vulnerable people," he said. "The price level for them is not only a question of higher prices. It becomes a question of survival, of just having access to food."[17]

Despite all the views quoted here, the majority of the world's policy analysts, defense experts, and governments persist in viewing famine as a consequence of war—not war as a consequence of famine. This monocular perspective creates dangerous international blind spots both for peace and for hunger.

WATER WARS

In 2007, the Egyptian-born World Bank vice president for the environment Ismail Serageldin warned, "Many of the wars of the 20th century were about oil, but wars of the 21st century will be over water." Former UN secretary general Boutros Boutros-Ghali warned bluntly that the next Middle East war might well be over water. A decade later his successor, Ban Ki-moon, was pressing home the same message at the World Economic Forum with an even greater sense of urgency: "Our experiences tell us that environmental stress, due to lack of water, may lead to conflict, and would be greater in poor nations. Ten years ago, even five years ago, few people paid much attention to the arid regions of western Sudan. Not many noticed when fighting broke out between farmers and herders, after the rains failed and water became scarce. Today, everyone knows Darfur. More than 200,000 people have died. Several million have fled their homes," he said.[18]

That disputes over water can lead to war should hardly be news. The former Israeli prime minister Ariel Sharon once told interviewers that water was one of the drivers of the 1967 Six-Day War. Attempts by each side to lay claim to water that the other regarded as theirs triggered armed border clashes feeding directly into events that led to the war.[19]

Studies by water policy analysts at the Pacific Institute indicate that the regions of the world most affected by conflict over water in recent years were the Middle East, Africa, and Central Asia. However, Latin America, China, India, and many other Asian countries have also experienced outbursts of violence over water supplies. Many countries have seen disputes, sometimes violent, between farmers and industry or urban water users. More nations affected by water disputes are now nuclear-armed: India and Pakistan face one another across the contentious waters and food bowl of the Indus. And cyber-terrorists seem to have developed a predilection for computer attacks on water agencies.[20]

In East Asia, Alan Dupont thinks that as economic and political interdependence grows between states, the water problems of one will spill over to become the problems of the others—and foresees intensifying disputes over common resources such as the Mekong River or simply over access to freshwater for the region's swelling megacities.[21]

The number of actual wars for which water was a key precipitating factor is hazy. A study by a graduate student at Oregon State University concludes that of 1,831 water-related "events" involving 124 countries, almost one-third, 507, were "conflicts"—ranging on a scale from harsh words to flying bullets—and the other two-thirds revealed a gratifying degree of cooperation in solving the problem. Although recently no war has been fought for water alone, the thirty-seven worst fights involving water tended to be sparked either by an acute regional shortage or by someone building a very large dam without seeking the downstream neighbors' approval. On the whole, the study suggests, countries still prefer to negotiate and collaborate rather than fight over water, and this fact deserves wider recognition. As global water scarcity increases and the climate becomes more unpredictable, however, so too does the scope for trouble.[22]

A long-running study compiled by Peter Gleick at the Pacific Institute lists thirty-five significant conflicts over water, mostly involving violence or terrorism, in the 1990s and forty-nine incidents in the period 2000–2007, which suggests a ratcheting-up in global water tensions.[23]

FISH WARS

Alan Dupont also highlights a category of international conflict that is taking place constantly under our noses and mostly beneath the radar of the global media: gunboat battles over ocean resources involving the navies of nations theoretically at peace with one another. These are minor flare-ups usually involving firing of live ammunition, boats being boarded and confiscated, and occasional bloodshed. Most go unremarked. "Fish is the main source of protein for an estimated 1 billion Asians, and fishing supports more people in East Asia than in any other region of the world. Over half the world's fish catch is taken in Asian waters," he explains. "Unfortunately, the Pacific is showing signs of environmental degradation from coastal pollution, overfishing and unsustainable exploitation of other forms of living marine resources. Asia has already lost half its fish stocks. . . . [A]s traditional fishing grounds are exhausted, competition for remaining stocks has intensified."[24] Fish wars, in other words, are an emerging form of eco-war.

Dupont notes growing friction, from the 1980s on, among the fishermen of various Asian countries, many of whom were arming themselves with machine guns and rocket-propelled grenades as a way of discussing their differences. By the late 1990s the death toll resulting from clashes between fishermen from one nation and navy patrol vessels from another nation was significant. He studied fourteen clashes—firefights and vessel seizures involving the navies of Thailand, Burma, Vietnam, Malaysia, China, Indonesia, the Philippines, North Korea, South Korea, Russia, and Japan—over marine resources, both fish and oil. "Interstate confrontation over fish and other marine living resources is emerging as a significant long-term security issue for East Asia," Dupont concludes. "The declining availability of fish is a global problem but East Asia's dependence on the oceans for food suggests that disputes over fish may trigger wider conflicts between regional states."[25] His Food Security Program at Sydney University is accumulating fresh evidence of a continuing buildup in these unseen maritime confrontations.

"Conflicts over the right to fish and to the fisheries resource are endemic in fishing industries all over the world, with some of these conflicts developing into open wars," warn Meryl Williams, a former director general of the scientific agency WorldFish, and her coauthor Choo Poh Sze.[26] Ethnic and national differences, the global rise of "pirate" fishing, and dwindling fish stocks at a time of soaring demand are all factors in disputes that sometimes involve even developed nations such as Britain and Iceland, or France and Spain.

Dupont considers major wars over fish to be unlikely, but he believes that as world fish catches dwindle due to overfishing, many states may come to regard them in the same light as oil and gas—resources worth contesting and defending by military force. Insignificant in themselves, these tiny sparks flicker at the edges of the tinder of larger regional food shortages that will emerge as a consequence of the coming famine. Far from comedic episodes, fish wars are a foretaste of what is to come as resources run low.

REFUGEE TSUNAMIS

"Internal wars lead to the displacement of enormous numbers of non-combatants, whose only option is to escape the violence and find refuge," the Peace Research Institute of Oslo pointed out in its seminal study on food and war. "People flee across immediate borders, sometimes destabilizing entire regions, leading to more conflict and more refugees."[27]

After a lull in the early years of the twenty-first century, the world refugee population began to climb again, reaching 42 million in both 2007 and 2008, according to the United Nations High Commissioner for Refugees (UNHCR). Of these refugees, the UNHCR found, 26 million were fleeing armed conflict.[28]

More graphically perhaps, the number of refugees roaming the Earth in search of peace, security, and sustenance represents a nation of the dispossessed as large as Spain—yet a nation with a difference: four-fifths of its citizens are women and children.

The term *refugee* has a rather specific meaning in international law and bureaucracy. It means a person who, "owing to a well-founded fear of being persecuted for reasons of race, religion, nationality, membership of a particular social group, or political opinion, is outside the country of their nationality, and is unable to or, owing to such fear, is unwilling to avail him/herself of the protection of that country."[29] The definition was later expanded to include people fleeing war, violence, and natural disasters, but the term still is not commonly applied to people fleeing hunger. The modern approach to refugees was originally adopted to cope with the 1.5 million who fled from Russia during the 1918–26 revolution and civil war, themselves events ignited and propelled by famine. War caused by famine no longer features in the common definition, however, and this has possibly muted awareness of the impact of hunger on global refugee movements and their wider consequences.

Most of today's refugees emanate from the strife-torn regions of Africa, the Middle East, and Central and South Asia, and the vast majority of them tend simply to flee to neighboring countries or other provinces of their own. Contrary to media depictions of an "invasion" of the wealthy countries, four out of five refugees remain within their own region—although very often this throws an intolerable burden on neighboring countries that have few resources of their own to look after them. Indeed, in many cases, it can lead to low-level conflict and even war between the incoming people and local residents, chiefly over access to land and water. For example, low-level conflict raged for many years between contending groups in the Indian states of Assam and Tripura following the arrival of more than ten million Bangladeshi refugees fleeing hunger in the 1970s and 1980s. NATO's involvement in the 1990s war in the Balkans was motivated in part by a desire to prevent the conflict from widening and precipitating a refugee flood into the rest of Europe.

Three elements have changed in recent times, however. First, a good half of all refugees, even if they originate in rural areas, now head for cities as their haven—which in turn puts great indirect pressure on the land and water resources that support the city and hence on the local farmers who feed it. Second, the number of emigrants from the moneyed and educated classes of countries facing scarcity of food, land, and water and potential instability has risen sharply since the start of the twenty-first century. Canada has regularly accepted close to a quarter million migrants each year since 2000 and the United States around a million.[30] About 200 million people around the world each year foresee trouble brewing in their homelands—including conflicts over food, land, and water—and are moving with their families to avoid it. The refugee wave, in other words, is often preceded by a far more orderly tide of far-sighted emigrants.

The third thing that has changed is the media. When the great famines struck Russia and China during the middle of the twentieth century, there were not only armed guards between the refugees and escape but also, thanks to censorship, a generally low awareness of the prosperity and personal opportunities afforded by the outside world. Many people died where they lived rather than risking the unknown. This no longer holds true: today television, magazines, and the Internet are bursting with the charms of affluence, the prodigal lives of celebrities, and the houses, luxury cars, consumer goods, and sumptuous meals that one-third of humanity enjoys; these media are ubiquitous and have penetrated to even the remotest corners of the world. Potential refugees now know, at some level, that there is a far better life to be had elsewhere if they have the strength, courage, and means to reach for it. Abetted by modern transportation and people smugglers, they now flee farther and more swiftly than ever to obtain it.

In 1845–51 Ireland's staple potato crop was blighted, casting the country into starvation and misery. The nation was critically dependent on the potato because in Ireland a farmer could grow three times as much food from potatoes as from grain, from the same area of land. When the ensuing famine ended after ten years, 750,000 Irish had died and two million had emigrated to the United States, Canada, and Great Britain.[31]

The Great Irish Famine may belong to the nineteenth century, but it carries undeniable messages for the twenty-first. It reveals the effect of overreliance on a major food source, and even though few countries today are as dependent on a single crop, many crops around the world are vulnerable to the loss of critical inputs such as water, fuel, or fertilizer, to disease, or to weather disaster. Any of these can unleash a local or regional famine—and the famine, in turn, will release an outpouring of refugees. In a world as heavily populated as ours, it is not hard to imagine how refugee tsunamis could result from a general food failure in the Indian subcontinent, Central Asia, China, sub-Saharan or North Africa, or Southeast or East Asia.

Events of this scale are beyond all previous human experience for the simple reason that the world has never been so populous or its resources so fragile. The possibility of regional crises involving twenty, fifty, even as many as two hundred or three hundred million refugees must now be seriously contemplated. Such floods are unlikely to be stemmed by military force. They will alter the politics, demography, and culture of entire regions. They will change history.

This is the most likely means by which the coming famine will affect all citizens of Earth, both through the direct consequences of refugee floods for receiving countries and through the effect on global food prices and the cost to public revenues of redressing the problem. Coupled with this is the risk of wars breaking out over local disputes about food, land, and water and the dangers that the major military powers may be sucked into these vortices, that smaller nations newly nuclear-armed may become embroiled, and that shock waves propagated by these conflicts will jar the global economy and disrupt trade, sending food prices into a fresh spiral.

Indeed, an increasingly credible scenario for World War III is not so much a confrontation of superpowers and their allies as a festering, self-perpetuating chain of resource conflicts driven by the widening gap between food and energy supplies and peoples' need to secure them.

RISING THREATS

Round the world, defense departments are already planning for what they anticipate as an era of rising instability and threats as populations swell, resources become scarcer, and climatic impacts hit home. Among the most notable examples is a U.K. Ministry of Defence *Strategic Trends* study that, among other insightful predictions, anticipated the collapse of global financial markets and the U.S. stock market by almost two years. Relevant findings from this report include:

- increased risk of food price spikes and shortages,
- water scarcities contributing to tensions in already volatile regions,
- mass population displacement due to climate or resource scarcities,
- possible collapse in fish stocks,
- increased risk of development failure in some countries and "megacity failure," and
- greater societal conflict involving civil war, intercommunal violence, insurgency, pervasive criminality, and widespread disorder (see map 2).[32]

Another important report, this time from a U.S. perspective, is "The Age of Consequences." This study explores the risks of a similarly destabilized world, erupting out of three different possible scenarios for climate change. Under the conservative scenario envisaged by the Intergovernmental Panel on Climate Change, this report anticipates "heightened internal and cross-border tensions caused by large-scale migrations; conflict sparked by resource scarcity, particularly in the weak and failing states of Africa." Under severe climate change, it foresees that "the internal cohesion of nations will be under great stress . . . both as a result of a dramatic rise in migration and changes in agricultural patterns and water availability. The flooding of coastal communities around the world has the potential to challenge regional and even national identities. Armed conflict between nations over resources, such as the Nile and its tributaries, is likely and nuclear war is possible."

The catastrophic scenario, the report simply says, "would pose almost inconceivable challenges as human society struggled to adapt," adding, "No precedent exists for a disaster of this magnitude—one that affects entire civilizations in multiple ways simultaneously."[33]

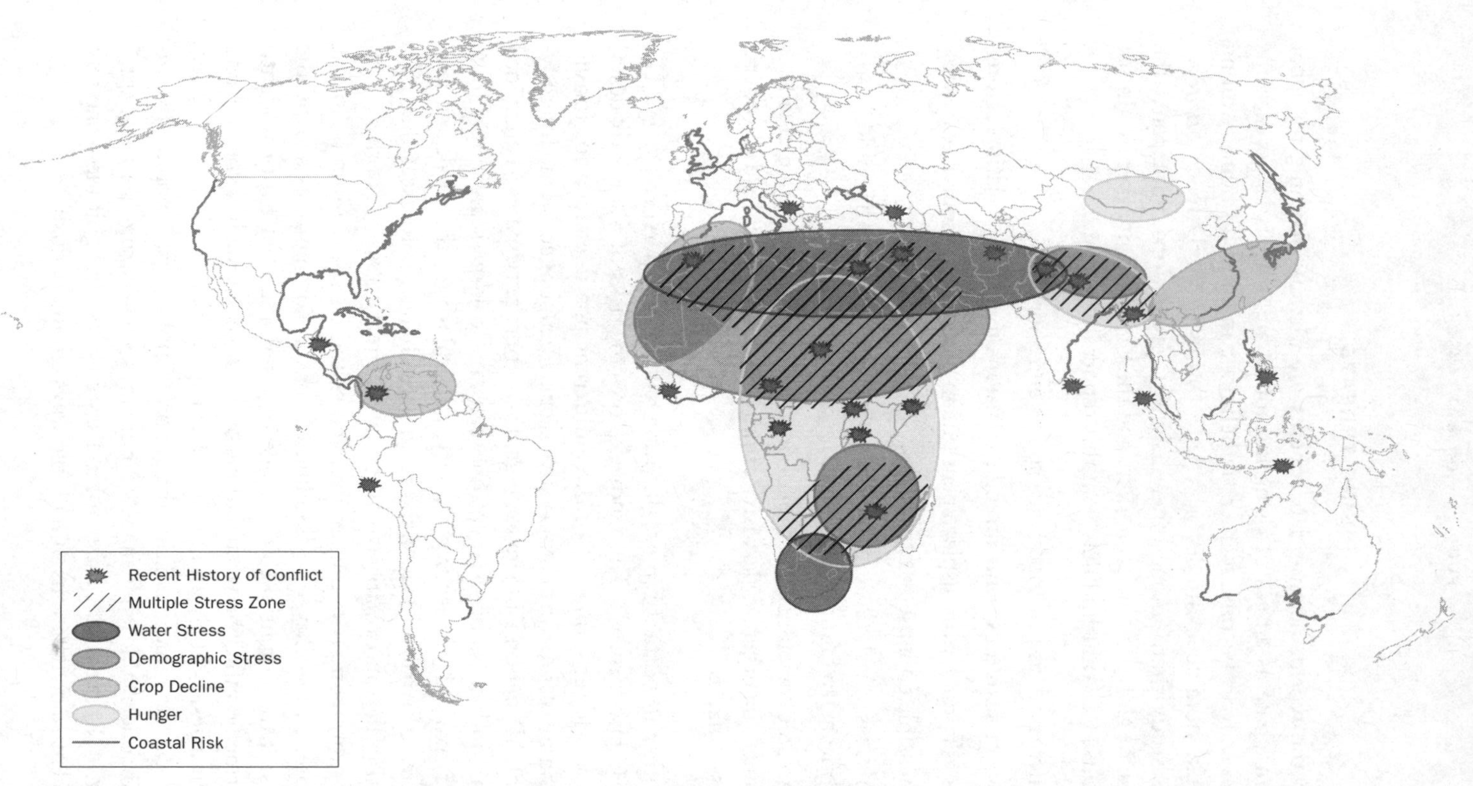

Map 2 Relationship between experience of environmental stresses and armed conflicts and refugee crises. British defense experts believe conflicts and refugee crises are most likely to break out in areas experiencing multiple stresses and scarcities. Source: Development, Concepts and Doctrine Centre, *The DCDC Global Strategic Trends Programme, 2007–2036* (London: U.K. Ministry of Defence, 2007), fig. 6, p. 69, www.dcdc-strategictrends.org.uk/viewdoc.aspx?doc=1.

AND NOW THE GOOD NEWS . . .

The good news in this grim picture is that, if shortages of food, land, and water spawn wars and refugee crises, the opposite also holds true—securing food for everyone can lessen or remove one of the primary tensions that lead to war and failed states, and ease the disruptions and shortages that prompt refugees to flee.

Achieving a secure global food supply is, in other words, an essential form of defense—although it is doubtful that any government on Earth perceives it quite that way. It is high time that they all did. Food, as Jimmy Carter pointed out, is a powerful peacekeeper.

Furthermore, a secure food supply is a way to prevent conflict that is often easier to attain than resolving intractable ethnic, political, or land-tenure quarrels: if the core of the local dispute is that not enough sustenance for the local population is being produced from the available land and water, then science and technology have demonstrated many times that they can supply the answers. Higher crop yields lower tensions. Even where strong differences remain, having enough food to feed the children may prevent the other issues in dispute from bursting into flame.

From our past and present successes, we also know that it is within our power to achieve complete food security, both locally and regionally. True, it is costly to achieve food and water security—but not nearly as costly as failing to achieve them. And whatever the cost, it is still only a fraction of the $1.5 trillion the world spends annually on armaments.[34] If the defense ministers of the world agreed to devote just 10 percent of their defense budgets each year to war prevention through food security, it could make all the difference.

The burgeoning demands of humanity, combined with the growing resource scarcities that undermine our every effort to achieve food security globally, make this—even more than global climate change—the greatest and most urgent challenge of our age.

In the twenty-first century, we either eat—or we fight.

THREE

THE WELL RUNS DRY

"If we don't get three inches, man,
Or four to break this drought,
We'll all be rooned," said Hanrahan,
"Before the year is out."

—J. O'Brien, *Said Hanrahan*

The farmers and fishermen of Lake Bam are watching as catastrophe unfolds before their eyes: their lake, which has sustained them for centuries, is slowly disappearing. Lake Bam lies in the sub-Saharan country of Burkina Faso, near the small city of Kongoussi. Its broad, shallow waters yield fish and water for drinking, livestock, and irrigated crops and create a pleasant microclimate not far from the Sahara desert. More than sixty thousand people rely on it for livelihood and survival. Yet, like innumerable lakes, river basins, and aquifers all around the world, Lake Bam is drying up.

The lake, which is part of the Volta river system in West Africa, fluctuates naturally with the seasons, expanding when the rains replenish it and contracting as the fierce heat of summer drinks up to two-thirds of its shallow waters. Of late, however, the rains have been sparse and the lake has dwindled even more severely in summer, fragmenting into a chain of muddy ponds less than a meter deep. Soil eroded by farming and grazing in the surrounding catchment is filling it in, too much water has been taken from its feeder streams, and the local climate appears to be drying out. The waters once supported nearly a thousand hectares of irrigated crops, but these have dwindled to one hundred and fifty. The nets of fishermen often come up empty. Yet the main sources of income in this, one of the world's most impoverished regions, are crops and fish.[1]

On one level, Lake Bam's story is that of the immense challenges facing Africa and its people: its far larger neighbor, Lake Chad, which supports thirty million people, shrank to less than one-tenth of its former size over thirty years, and is forecasted to disappear completely by 2030. On another level, it illustrates what is happening to water around the world and foreshadows events that will affect every one of us as the century advances. In Hubei, China, once known as the "Province of a Thousand Lakes," 815 lakes are said to have dried up by 2001.[2]

"This year, the world and, in particular, developing countries and the poor have been hit by both food and energy crises," Colin Chartres, head of the nonprofit International Water Management Institute (IWMI), declared in 2008. He continued,

> As a consequence, prices for many staple foods have risen by up to 100%. When we examine the causes of the food crisis, a growing population, changes in trade patterns, urbanization, dietary changes, biofuel production, and climate change and regional droughts are all responsible. Thus we have a classic increase in prices due to high demand and low supply. However, few commentators specifically mention the declining availability of water that is needed to grow irrigated and dryland crops. According to some, the often mooted solution to the food crisis lies in plant breeding that produces the ultimate high yielding, low water-consuming crops. While this solution is important, it will fail unless attention is paid to where the water for all food, fibre and energy crops is going to come from.[3]

A farmer who faces a critical shortage of diesel fuel can resort to growing his own fuel crops or even use draft animals. But every form of food production, whether crop or livestock, on the farm or in the factory, depends on water. No water, no food.

Food production is a thirsty business. Today food grown by irrigation uses about 70 percent of Earth's readily available freshwater. The water we use in our homes and towns takes up 10 percent, while power generation and industry consume the remaining 20 percent. This 70:30 balance between rural and city use, however, is now changing as cities expand—and begin to swallow more and more of the farmers' scarce water. Urban demand for water may soar by as much as 150 percent by 2025 as the cities themselves burgeon.[4]

Irrigation has been practiced for at least eight thousand years and is a cornerstone of civilization because it enables one person to produce food for many. This underpins the growth of cities and industries. Irrigation has the great advantage over dryland farming that it can yield crops or pastures more reliably, especially in regions where rainfall is erratic. Today dryland farming uses about 5,000 cubic kilometers (1,219 cubic miles) of rainwater, while irrigated food production uses around 2,700 (658 cubic miles). Irrigation supplies 45 percent of the world's food, however—and we may well need to increase this to 60 percent in order to feed eight billion people by 2025 and more beyond.[5]

The world is running out of water readily available to do this—at a time when global demand for food is set to double.

WATER EATERS

We "eat" an awful lot of water. The average calorie we put in our mouths takes a liter of water to produce. People around the world consume between 1,800 and 3,900 calories in food energy per day, on average, so those on affluent diets go through at least 3 tonnes (792 U.S. gallons) of water every day—around 1,240 tonnes (327,000 gallons) per year. This compares with the two or three liters (two-thirds of a U.S. gallon) a day we need for drinking. Since life expectancy in developed countries is now mostly around eighty years, this means that the average affluent person "consumes" nearly enough water in a lifetime in the form of food to float the USS *Enterprise,* a rather large aircraft carrier that displaces ninety thousand U.S. tons.[6] Just over half of this water comes directly from rainfall. The rest comes from irrigation, drawn from limited resources such as rivers, lakes, and groundwater aquifers, which it is possible to exhaust by taking out water faster than it is naturally replenished by rainfall, runoff, or underground flows.

Most people are unaware of how much water goes into producing the food on their plates—yet this is likely to be a critical factor on which we will all base our food choices in the years ahead, as water becomes scarcer and more costly. Food uses so much water because the plants on which all our foods are based—including livestock products—take up and release a great deal of water as part of their internal functioning, just as we breathe air in and out. This process is known as transpiration. The water embodied in our food is referred to as "virtual water."

To get an idea of the virtual water content of our food and other products, see table 1.

Using a different method of estimating water content under Australian conditions, the irrigation scientist Wayne Meyer calculates that

Table 1 WATER REQUIRED TO PRODUCE COMMON FOODS AND PRODUCTS

Food or Product	Water (in liters)	Water (in gallons)
Slice of bread	40	11
Potato	25	7
Tomato	13	3
Cup of coffee	140	37
Glass of milk	200	53
Egg	135	36
Glass of wine	120	32
Kilogram of grain	1,500	396
Liter of palm oil	2,000	528
Kilogram of chicken	6,000	1,585
Kilogram of beef	15,000	3,962
Hamburger	2,400	634
Cotton T-shirt	4,000	1,057
Pair of leather shoes	8,000	2,113

SOURCE: Lenntech, "Use of Water in Food and Agriculture," Lenntech.com, n.d., www.lenntech.com/water-food-agriculture.htm. There are numerous estimates of how much water it takes to produce food. These figures were compiled by Lenntech, a water technology company originating from the University of Delft, The Netherlands.

- producing 1 kilogram of oven dry wheat grain takes 715–50 liters of water (189–98 U.S. gallons),
- 1 kg maize takes 540–630 liters (143–66 U.S. gallons),
- 1 kg soybeans take 1,650–2,200 liters (436–581 U.S. gallons),
- 1 kg paddy rice takes 1,550 liters (409 U.S. gallons),
- 1 kg beef takes 50,000–100,000 liters (13,200–26,400 U.S. gallons), and
- 1 kg clean wool takes 170,000 liters (45,200 U.S. gallons).

So that smart wool suit you bought took at least 170 tonnes (45,200 U.S. gallons) of water to grow the grass that fed the sheep that grew the wool that you wear.[7]

From this it can be seen that the average well-off person easily manages to use from two to three tonnes (2.2–3.3 U.S. tons) of water, in the form of food or fiber, every day—an Olympic-sized swimming pool of water every two and a half years. Also, certain diets, especially those rich in meat, dairy, or oils, use massively more water than diets based mainly on vegetables or grains. Food and fiber thus account for seven times the amount of water we use to drink, wash, clean our homes, flush toilets, or water parks, gardens, and sports fields. This is arguably our biggest personal impact on the planet.[8]

HOW MUCH FRESHWATER IS THERE?

Superficially, the world has an abundance of freshwater—but it isn't always located where the people are or where the food is grown, a lot of it is frozen, it isn't always easy or cheap to extract, and often it is too polluted to use. We have already tapped most of the freshwater that is accessible and economic, and tapping more distant, dirty, or difficult supplies of water, like oil, is a great deal more costly. At the same time, we manage, conserve, and price water very badly.

World water figures should be regarded with some caution: for many countries the data are poor or out of date. The broad scientific estimate of world freshwater and where it goes, however, can be seen in table 2. Out of all this water, humanity actually extracts about 3,900 cubic kilometers (853 cubic miles) for all of its uses, of which 2,700 cubic kilometers (560 cubic miles) are employed to irrigate crops.[9] This is what is meant by "available water," the amount we can readily and affordably harvest from rainfall, rivers, lakes, and aquifers on an annual basis. This is the water that is in increasingly short supply because we have overextracted it from most rivers, and their flows are now no longer adequate for a healthy environment, either on land or in the sea.

WET FEET

Our use of water is known as our water footprint. According to the UNESCO Institute for Water Education, humanity's global water footprint is 7,450 billion metric tonnes (7,450 cubic kilometers or 1,817 cubic miles), which works out to 1,243 metric tonnes (327,000 U.S. gallons) per person per year. This footprint varies considerably from country to country, however: the average American, for example, has a water footprint of 2,483 tonnes (655,000 U.S. gallons) per year whereas the average Chinese has one of 702 tonnes (185,000 U.S. gallons)—part of the difference being due to meat consumption. Countries that are warm and have high evaporation rates use more water to grow food, and thus have large water footprints—examples being Italy and Greece, which both use around 2,300 tonnes per capita annually. Countries whose climate

Table 2 SOURCE AND DISPOSITION OF WORLD'S FRESHWATER

	Cubic kilometers	Cubic miles	Percentage of world's total freshwater
Source of freshwater			
World rainfall	110,000	27,000	100
Disposition of freshwater			
Evaporation from landscape	62,000	15,000	56
Transpiration from dryland agriculture	5,000	1,220	5
Withdrawal for irrigated agriculture	2,700	560	3
Loss from storage	1,500	366	1
Withdrawal for use by cities and industry	1,200	293	1
Runoff into sea	38,000	927	35

SOURCE: Based on International Water Management Institute, summary of *Water for Food; Water for Life: A Comprehensive Assessment of Water Management in Agriculture,* ed. David Molden (Sterling, Va.: Earthscan, 2007), box 1, p. 6, www.iwmi.cgiar.org/Assessment/files_new/synthesis/Summary_Synthesis Book.pdf.

favors efficient food production tend to have lower water use per capita, such as the Netherlands (1,220 tonnes per capita) or Australia (1,390 tonnes per capita). Around one-sixth of the world's accessible water is used to grow food for export to other countries, making water-poor countries increasingly reliant on those with ample supplies.[10]

WHY IS WATER BECOMING SCARCE?

Today our freshwater supply is being stretched to the limit. The main reasons include the following.

Food demand growth. Since 1950 the area of irrigated land has doubled and withdrawals of water have tripled due to our growing population and even stronger demand for high-protein food. It is conservatively estimated that an additional 6,000 cubic kilometers (or 1,500 cubic miles) of freshwater will be needed for irrigation to meet future increases in global population and food demand.[11]

City growth. Worldwide, 3.5 billion people now live in cities. By 2050, it is quite likely that cities will have 7 billion inhabitants. Current urban water demand of 1,200 cubic kilometers (300 cubic miles) thus could balloon to around 2,500 cubic kilometers (600 cubic miles) or even more.[12] As cities are usually richer than farmers, they can afford to buy water once used to grow food and divert it to urban uses, which tends to reduce local food production and cause food prices to rise.

Economic growth. As populations develop economically, their per capita consumption of meat, dairy, sugar, and oils rises, and these foodstuffs require far more water to produce than vegetables and grains do. The more successful we are at overcoming poverty and building our economies, the heavier the demands we place on water resources.

Overextraction. Mechanical pumps powered by fossil fuels allow the extraction of surface and underground water in immense volumes, which often exceed natural rates of recharge. This is causing the level of groundwater to fall rapidly in most countries where it is used to grow food. Pumping of groundwater also helps to empty rivers, lakes, and wetlands and kills landscapes when water tables sink out of the reach of tree roots. In many countries, governments encourage this destruction and waste by providing cheap hydroelectricity or by underpricing water.

Ignorance. Most communities and many water authorities do not in fact know the true extent of their water resource or the rate at which it is replenished. This makes every decision to use it a gamble—and one that often goes wrong, resulting in a dying river, lake, aquifer, basin, or even sea.

Confusion. Many people, including governments, seem unaware that surface water in rivers and lakes and groundwater drawn from wells and aquifers are often interconnected and that removing one reduces the supply of the other. People often regard well water as a "free good." Consequently, more water may be withdrawn than the total system receives in recharge.

Contamination. Water discharged from farms is frequently muddy and contaminated with nutrients and chemicals, making it unsuitable for other uses such as growing fish or drinking. Water used in cities is usually contaminated by industry with heavy metals or toxic organic pollutants, or else contains sewage (treated or not), oils, chemicals, animal excrement, fertilizer, and many wastes, making it unsuitable for reuse unless cleansed. An estimated 1.5 billion people drink contaminated water, which kills 2–5 million every year.[13] Pollution is a major factor in the global scarcity of clean freshwater.

Poor management. Most of the world's water is managed badly or not at all. There are fierce disagreements in many communities over water rights and competing uses. Authorities often sell (or permit the removal of) more water than enters the river naturally from rainfall and recharge, causing river basins to dry up. Water management in many countries has failed to recognize the new global realities of water scarcity or adapt with the times. Water policy is often poor and water management agencies frequently lack transparency, independence, expertise, and accurate information about the size of their resource.

Economics. Water may be in short supply because some countries simply cannot afford to develop their resources fully or to manage them well. In other cases, distorted water pricing sends the wrong signals to users and causes waste or overuse. In some countries, the availability of cheap, subsidized electricity has led to overpumping of water. In others, high energy costs make water unaffordable to the poor.

Poor farming practices. Very large amounts of water are wasted by inefficient farming, growing unsuitable crops, and so on. The most efficient farmers can grow the same amount of food using only a fraction of the water used by their less efficient fellow producers—so there is plenty of room for improvement.

Poor infrastructure. Dams, channels, pipes, and storage containers are in poor shape in most countries and lose huge amounts of water through leakage, evaporation, theft, and so on.

Desertification. In places where landscapes have been heavily overcleared or overgrazed, there is evidence of declining natural rainfall, caused by the removal of vegetation and the loss of its contribution to the local hydrological cycle.

Reforestation. In countries where forests and natural landscapes are being restored, water flows into rivers are reduced by the natural pumping action of trees and plants.

Salt and acidity. In drier regions the removal of natural vegetation can often lead to the landscape becoming salty and unproductive due to the rise of saline groundwater. Where sediments rich in sulphides are dried out by the extraction of groundwater, deadly acid can flow into rivers, wetlands, and estuaries, killing fish and water plants.

Natural toxins. Arsenic, fluorine, and other naturally occurring toxic compounds may be mobilized by groundwater extraction. Dissolved arsenic in drinking and cooking water poisons an estimated one hundred million people worldwide and is causing an epidemic of cancers.

Development focus. Aid programs have naturally focused on providing clean water for water supply and sanitation, and the need to improve the management of the water resource itself has been relatively neglected.

Governments also have focused on meeting urban water needs at the expense of agricultural water.

Climate change. Key regions of the world are starting to dry out, in particular the grain bowls, while others may be getting wetter. Irrigation regions that rely on glacial or snow melt are particularly at risk. Increasing pressure on governments to tackle carbon emissions to combat climate change has led to reduced priority for many water-management improvements.

"Fifty years ago the world had fewer than half as many people as it has today. They were not as wealthy. They consumed fewer calories, ate less meat, and thus required less water to produce their food. The pressure they inflicted on the environment was lower. They took from our rivers a third of the water that we take now," explains IWMI, the world's leading water research center.[14]

In the last four decades of the twentieth century, the amount of freshwater available for each human being worldwide shrank by almost two-thirds. It is expected to be halved again by 2025.[15] Such stark numbers have prompted a number of observers to assert that the world has already passed "peak water" and that supply is now declining, both per capita and relative to demand. This poses a serious hazard to our ability to maintain, let alone double, world food supplies.

Water shortages, no matter where they occur, will be felt in the wallet of every person who buys food. Well before the spike in global food prices, Joachim von Braun and his colleagues at the International Food Policy Research Institute warned that an international water crisis could send prices through the roof: "Insufficient attention to water-related investments and policies could produce a water crisis that would in turn lead to food system stress, given competing demands on scarce water. With increased water stress, relative crop yields decline, representing an annual loss in crop yields forgone," they said. "In such a water crisis scenario, cereal production declines by 10 percent, a loss equivalent to the entire Indian cereal crop. This decline would cause rice prices to rise by 40 percent, wheat prices by 80 percent, and maize prices by 120 percent by the year 2025. Price increases of this magnitude will dampen demand, contract trade, and hit poor people the hardest, especially the 1 billion people who live in urban slums and the many millions of rural poor people who are net purchasers of food."[16] The impact of water shortages will not be confined to countries where supplies are short: through their effect on food prices, virtual water, and world trade, water shortages will affect almost everyone.

Such warnings have been sounded since the early 1990s by the Food and Agriculture Organization (FAO) and others—in statements such as, "Human demands are about to collide with the ability of the hydrological cycle to supply water"—and they ought to come as no surprise.[17] Governments have also been warned that large parts of the world are running into water scarcity, ranging from chronic to acute. These tend to fall along the tropics of Capricorn and Cancer, bands of the planet that are naturally dry and are where most of the world's deserts occur. They are also regions that are already highly populated and where both population and food demand are rising rapidly. About one person in four lives in a region of the world where water supplies are under stress. Map 3 gives a general impression of where looming shortages of water are likely to occur.

The potential impact on individual countries may be illustrated by the case of China—a land with 22 percent of the world's people and just 8 percent of its available freshwater. The Chinese Ministry of Water Resources has warned of a serious crisis by 2030 due to falling per capita water availability.[18] More than half of China's 660 cities experience regular water shortages, and 70–75 percent of the water used to supply the nation's main food bowl, the North China Plain, comes from groundwater and is shrinking. As much as 90 percent of China's urban water is said to be polluted. "China's water crisis threatens global prosperity and stability," says the water expert John McAlister, who adds that availability in key parts of China is already well below the danger level for economic and social disruption. Before he became premier, Wen Jiabao was quoted by the *China Daily* as saying, "The shortage of water threatens the survival of China." The nation's first environment protection minister, Qu Geping, told media that China's water resources would support only a population of 650 million—yet China's population is expected to pass 1.6 billion. China's planners are keenly aware of the jeopardy their country faces: a network of mighty canals is urgently being built along three main routes to shift supplies from the south to the north, but whether such engineering solutions can succeed remains to be seen.[19]

As to how water scarcities will play out for humanity in the next two generations, IWMI projections indicate that most humans are now treading a path similar to China's. By 2050, they say, more than two billion people will face severe scarcity and an additional five billion will experience moderate scarcity. "The causes of water scarcity are essentially identical to those of the food crisis. There are serious and extremely worrying factors that indicate water supplies are steadily being used up,"

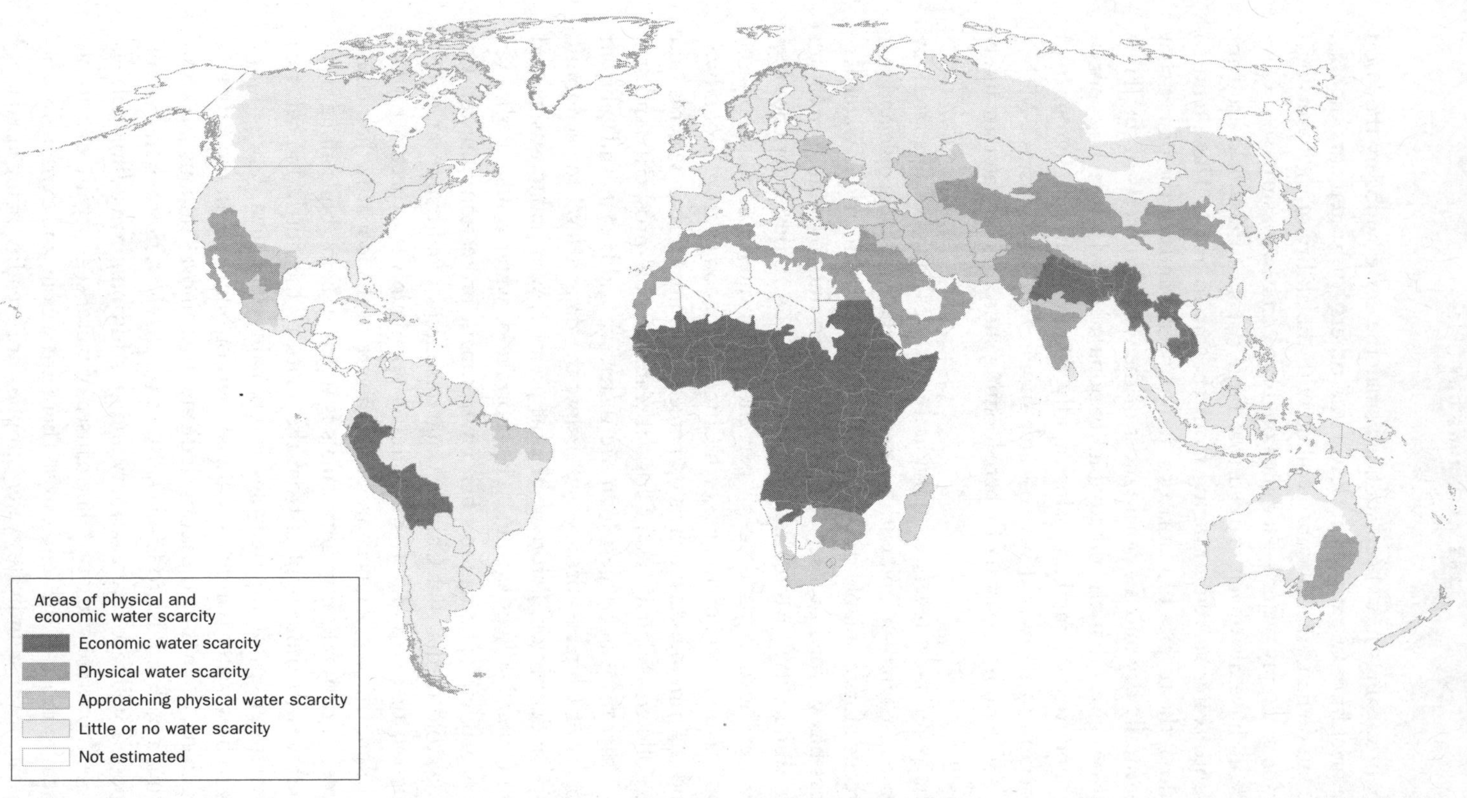

Map 3 Areas of looming water scarcity for human uses. Source: International Water Management Institute, summary of *Water for Food; Water for Life: A Comprehensive Assessment of Water Management in Agriculture,* ed. David Molden (Sterling, Va.: Earthscan, 2007), map 2, p. 11, www.iwmi.cgiar.org/Assessment/files_new/synthesis/Summary_SynthesisBook.pdf.

warns the IWMI. "Given that one liter (more for livestock products) is used to produce one calorie of food and taking food losses into account, it will take up to 6000 cubic kilometres of additional water annually to feed another 2.5 billion people 2,500 calories per day. This is almost twice what we use today and is not sustainable," adds IWMI director general Colin Chartres. "A roll call of countries where water scarcity already undermines food security includes China, India, Pakistan, most of North Africa and the Middle East, and large parts of southern Africa. In the coming decades as countries struggle to feed their growing populations, many will have to import large amounts of food, putting a major drain on their economies," he continued.[20]

The FAO expects global meat production alone to rise by 185 million tonnes (204 million U.S. tons) per year, which on its own would use 1,850 cubic kilometers (450 cubic miles) of water.[21] The IWMI concludes, bluntly, "Unless we change the way we use water and increase water productivity . . . we will not have enough water to feed the world's growing population."[22] The result, it says, will be water crises and possible conflicts in many parts of the world.

THE GROUNDWATER CRISIS

The state of the world's water supplies is highlighted by the case of groundwater, which is in a dire condition in almost every country where it is used to grow food. People everywhere have a disturbing tendency to view groundwater as an inexhaustible resource and to extract it far more quickly than it can recharge. Furthermore, many do not realize that much of the water they see in a river (especially during the dry season) is actually "baseflow" from groundwater that is connected to the river, rather than surface runoff from tributary streams—and that pumping groundwater even from wells many miles away can still cause the river to shrink or dry up.

Groundwater provides about one-fifth of humanity's freshwater. It consists of rainfall that percolates from the surface down into aquifers, where it is stored, free from evaporation. Depending on the silt, sand, and rock it has to pass through on its underground journey, groundwater can be days, years, or even millions of years old. Libya's Great Manmade River taps fossil water that accumulated beneath the Sahara forty thousand years ago—and is unlikely to be replenished anytime soon. This is known as "water mining," extracting water in the clear knowledge that it will eventually run out completely.[23]

In the United States, for instance, groundwater supplies more than half of all drinking water and more than one-third of all agricultural water needs. The huge Ogallala aquifer, which underlies eight states in the American Midwest and is extensively used to grow food, is being depleted at ten times the rate of natural recharge, and some experts fear it could dry up completely within twenty-five years. In places, the depletion rate was more than one hundred feet until 1980 and was more than forty feet between 1980 and 1997. This highlights a critical issue: once aquifers are pumped dry, they can collapse and then cannot be recharged ever again.[24] In India, groundwater tables have been dropping as much as three feet a year in many regions, forcing poor farmers who cannot afford gasoline-powered pumps to abandon irrigated food production. According to Neerjaal, a maker of software for monitoring and conserving groundwater, "Indian surveyors have divided the country into 5,723 geographic blocks. More than 1,000 are considered either overexploited, meaning more water is drawn on average than is replenished by rain, or critical, meaning they are dangerously close to it." A recent report found that groundwater exploitation was particularly severe in the vital food bowl states of Punjab, Haryana, Gujarat, Maharashtra, and Andhra Pradesh.[25]

Six hundred Chinese cities rely extensively on groundwater for their survival, but in four hundred of these the resource is dwindling measurably. Beneath Beijing, the *People's Daily* reported, water tables were sinking by 1.5 to 2 meters (5–7 feet) a year. The North China Plain, heartland of the nation's wheat and corn production, waters more than two-thirds of its crops from aquifers that have sunk by as much as 90 meters (290 feet) in recent years, a "potentially disastrous problem" that could see the water run out within decades. Compounding the challenge is the fact that urban groundwater in three hundred Chinese cities suffers from toxic industrial contamination. The groundwater problem is becoming more acute even in parts of the world where population is static: in 60 percent of European cities with populations greater than one hundred thousand, underground water is being extracted faster than it is recharged.[26]

WATER AND THE ENVIRONMENT

The impact of our water demands on the natural environment—the death of lakes, rivers, forests, wetlands, woodlands, and even seas and landscapes—is a storm warning to humanity of approaching scarcity. It also highlights another important but overlooked fact: that farmers are the managers not only of the food supply and the land they work but also of the water they use, water that is vital to many other human activities as well as to nature itself, both upstream and downstream of the farm.

Water is said to provide "ecosystem services," by which the experts mean keeping landscapes functioning so they can provide us with food, clothing, shelter, and clean water to drink as well as maintaining the landscapes, air, biota, and natural systems on which our survival ultimately depends. Healthy landscapes in turn provide fertile soil, soak up more carbon from the atmosphere, cleanse water, and maintain the diversity of life. Damaged landscapes spell the loss of these benefits. Poor management of agricultural water not only degrades the water itself for everyone and everything else that uses it but also has consequences for the wider landscape and the survival services it provides to humanity and to nature.

Scientists now urge the management of the "agro-ecosystem"—the landscape in which farming takes place, along with its resources and natural values—as a single whole. In practice this means that today's farmers are being asked not only to produce far more food (often for far less money)—but also to use less water, to clean it up, and to make sure it gets back into rivers, wetlands, and wilderness areas. This responsibility was highlighted by four hundred of the world's top agricultural scientists in a report to the World Bank in which they wrote of the need to "recognize farming communities, farm households, and farmers as producers *and* managers of ecosystems."[27] To do this, they add, there must be changes in incentives "all along the value chain," which is another way of saying that those who wish to eat in the future are going to have to start paying a lot more for the true environmental costs of producing their food, especially for the water used to produce that food.

WHAT ARE THE ANSWERS?

There are many ways in which the looming global water crisis can be alleviated and most are well within our technical abilities, though not necessarily within our governments' spending intentions or collective willpower. Actions to solve the world's water crisis may be divided into three broad areas: awareness, incentives (or penalties), and technical advances. To succeed, these have to be built into a framework comprising the essential policies, legislation, and institutional arrangements to make them work. The following list is far from complete but gives an idea of what is needed.

1. Awareness, education, and behavioral change
    - Change the way we think about water—from being a "free good" to a vital but scarce resource for which we all must pay.
    - Save water by reducing the current wastage of half of all food grown.
    - Educate consumers to make food, timber, and fiber purchases based on water saving.
    - Develop and promote low-water diets and farming systems.
    - Mandate recycling of urban water in order to reduce city calls on rural water.
    - Grow more food in cities, where surplus water is available for reuse.
    - Educate farmers, large and small, in the best water-saving techniques.
    - Open up food trade between water-rich and water-poor countries. Promote "water alliances."
    - Encourage all domestic and industry users to practice water thrift.
    - Invest vastly more in sharing knowledge about ways to save water among millions of farmers and water users globally.
    - Better manage the whole hydrological cycle—not just its parts—including rainfall, evaporation, overland flow, groundwater, wetlands, lakes, estuaries, and marine systems.
2. Incentives and penalties
    - Price water according to its real cost (including environmental and opportunity costs) to send the right signals to users.
    - Price food according to its water use and environmental impact; use the revenue to support greater on- and off-farm efficiencies and to reduce water use.
    - Introduce markets in water that allow it to flow to the most efficient and sustainable users.
    - Develop basin-level water plans that make it clear what water is available for consumption—and what must be reserved for the environment.

- Monitor all water extraction; penalize illegal extraction.
- End government subsidies that price water artificially low and encourage waste.
- If price signals prove ineffective, legislate to prevent land with good soil and reliable water from being diverted out of agricultural use and into nonessential water-wasting activities.

3. Technical solutions
    - Improve water-use efficiency on farms using a wide range of water-saving technologies, farming practices, and new water-efficient crop varieties.
    - Overhaul channels, dams, pipes, and other water infrastructure to reduce leaks and evaporation losses.
    - Expand rain-fed cropping where feasible.
    - Invest more in water science and advanced technologies that raise food and fiber output per unit of water.
    - Store and cleanse surplus water underground wherever possible (known as "aquifer recharge").
    - Recycle all urban water.
    - Develop novel food-production systems that use minimal water, such as biocultures.

Many books could be written about ways to save water and improve its use. This list serves simply to remind us that although the risks of a global water crisis are great and many, so too are the solutions and the opportunities.

Solving the global water crisis will involve substantial trade-offs. The FAO notes that in some regions, such as the Middle East, North Africa, and South Asia, water withdrawals already account for a large share of the total renewable water available. Other areas, such as Latin America, use only a small part of the water they could potentially harvest for food production—and therefore are viewed by some as having considerable potential as future food bowls. The drawback, however, is that converting these areas to arable farmland will entail the clearing of immense tracts of the Earth's dwindling forests and grasslands and the removal of more water from the environment, with consequences including massive loss of species, accelerated environmental degradation, and potentially major impacts on the global climate. Any area designated a future food bowl will need to have its water resources assessed with the greatest care, as well as all the competing demands for them—including the environment itself, fisheries, aquaculture, and cities.

This way of looking at the issue underscores the concept of virtual water, in which water-poor countries can import the shortfall in the form of food from countries with ample water. For this to happen, however, barriers to trade in food must be abolished, countries must learn to trust one another (more than they do today) to supply the basic necessities of life, and the food must be affordable to countries that are dry but poor.

Traditionally, food has been viewed by governments as an issue of national security, and they have used this to justify the maintenance of many inefficient and wasteful farming and water systems rather than importing food from the most efficient, competitive, and sustainable suppliers. In the twenty-first century, the world may no longer have that luxury: the time is coming when we will need to consider a global approach to food production, focusing such production wherever on the planet is most efficient and sustainable.

Although this approach may seem at first glance to be naive and politically impractical, the megacities offer hope that it can be achieved. Thanks to the rise of supermarkets and globalization of the food chain, most city people have little notion where their food actually comes from or how it is produced or processed, mainly caring that it is good, safe, plentiful, and cheap.

Addressing the crisis in world water and food security will require a mental shift away from past notions of local or national self-sufficiency—and the emergence of "global citizens" bold, wise, and generous enough to consign concepts such as national food security and free water to the dustbin of history in the interest of greater food security and geopolitical stability for everyone.

WHAT CAN I DO ABOUT IT?

The looming world water crisis won't be solved by governments and water authorities acting alone. They will need the wholehearted understanding, support, and behavioral change of entire communities that have learned to value their water as the primary ingredient of life and the keystone of their food security. Here are some practical suggestions for changes we can make in our own lives.

1. Practice water saving in our choice of food, drinks, and other purchases, which will send a message to farmers and manufacturers that we value products that save water.
2. Support realistic pricing of water, according to its true value.
3. Support moves for goods to be labeled and priced according to their virtual water content.
4. Support the recycling of urban wastewater and stormwater in order to reduce the pressure of urban demand on rural water supplies.
5. Support the separation of urban waste streams into recyclable/nonrecyclable and the replacement of the flushing toilet with composting or other systems.
6. Save water in our homes, gardens, and workplaces.
7. Avoid all waste of food.
8. Eat more vegetables and less meat, dairy, and oils: save the planet and avoid heart disease.
9. Grow more of our own food and support local food production.
10. Teach our children to prize water as much as freedom.

FOUR

PEAK LAND

> We owe a *cornfield* respect, not because of itself,
> but because it is food for mankind.
> **—Simone Weil, *The Need for Roots***

If people respected cornfields, as the French philosopher Simone Weil once suggested we should (as part of our love for our homeland), we would not build cities on them or degrade them. The coming famine of the midcentury is likely to teach us a renewed respect for grain fields, rice paddies, orchards, market gardens, and the soil that sustains them all.

Believe it or not, the world is running out of high-quality soil. In one sense, we passed "peak land" a long time ago. A report by Rabobank shows that the area of food production has declined from 0.45 hectare (1.1 acres) per person in the 1960s to 0.23 hectare (0.6 acre) currently and will keep on falling as population rises, to around 0.18 hectare (0.4 acre) in 2050.[1]

Another way to interpret this, however, is as a tribute to the remarkable achievements, over the past half century, of the world's farmers and agricultural scientists, who now put significantly more food on our plates using less land through advanced broadacre farming systems and efficient smallholder agriculture. It shows just what can be achieved when we put our minds to it.

This also underscores, however, that food security is a race—between the things we can do to increase it, such as using fertilizers, fossil fuels, better crop rotations, and improved varieties, and the things we do to destroy it, such as losing soil, water, and nutrients, exacerbated by our

Table 3 USAGE OF SUITABLE LAND FOR AGRICULTURE, BY REGION

Region	Cultivated area (millions of hectares)	Area suitable for agriculture (millions of hectares)	Percentage of suitable land in cultivation
Asia	439	585	75
Latin America	203	1,066	19
OECD	265	497	53
Russia	387	874	44
Sub-Saharan Africa	228	1,031	22
West Asia and North Africa	86	99	87
World	1,600	4,152	39

SOURCE: Ghislain de Marsily, "Water, Climate Change, Food and Population Growth," *Revue des Sciences de l'Eau* 21, no. 2 (2008): 111–28.
NOTE: 1 hectare = 2.47 acres

increasing population and demand for food. Our destiny depends on the state of this race in the middle of the twenty-first century.

Superficially, the world appears to have plenty of spare land, but—like water—it isn't always located in the most favorable climatic regions or where the major centers of population and increases in food demand are most likely to occur, as table 3 indicates.

This table shows that Asia, for example, has developed three-quarters of its available stock of land, whereas South America has developed only about one-fifth. To understand the significance of this contrast, we need to compare the available land area with the likely increase in food demand (table 4).

Combining the two tables shows that food demand in Asia is likely to more than double, but there is room only for a very modest 25 percent expansion in farmland. In West and North Africa the situation is more dire, with one and a half times more food required and only 13 percent more land to provide it. In Latin America the outlook appears more reassuring, with food demand likely to double but four times the current farmed land area available to satisfy it. In Russia and the member countries of the Organization for Economic Cooperation and Development (OECD), land supply is well ahead of likely increase in domestic food demand. In sub-Saharan Africa, the situation is delicately poised with a fourfold increase in food demand expected—and roughly four times the current farmed area available.[2]

Table 4 FOOD NEEDED IN 2000 AND 2050, BY REGION

Region	Cereals needed in 2000 (in millions of tonnes)	Cereals needed in 2050 (in millions of tonnes)	Rate of increase
Asia	1,800	4,150	2.34
Latin America	272	520	1.92
Sub-Saharan Africa	262	1,350	5.14
West Asia and North Africa	154	390	2.5

SOURCE: Ghislain de Marsily, "Water, Climate Change, Food and Population Growth," *Revue des Sciences de l'Eau* 21, no. 2 (2008): 111–28.
NOTE: 1 tonne = 1.1 U.S. ton

These bald numbers hide unpalatable truths, however, including, first, the fact that much of this "new" land is unlikely ever to be farmed because of the environmental destruction and loss of landscape function it would cause. Second, we are losing productive land through accelerated soil degradation faster than we can open up new areas. Third, because much of this new land has poor soils, it will require massive inputs of fertilizer and energy, or drainage, to make it productive. These issues lend weight to a view that humanity has already reached, or has passed, "peak land."

Another way to look at the issue is to chart growth in world population and food intake against expansion in the world's food-growing area (table 5). It can be seen that while the total area of land that is cropped or grazed worldwide expanded by 1.8 percent in the fifteen years between 1990 and 2005, this is substantially slower than the rate of growth in food consumption per person and a staggering twelve times less than the growth in population. Though these figures vary from year to year as land is taken in and out of production, the arable area, which produces most of the grain, oilseeds, fruit, and vegetables we eat or feed to livestock, is growing at only one-seventh the rate of consumption and one-forty-sixth the rate of population. Thus, from 1990 to 2005, world demand for food grew fifteen times faster than the area of land available to produce it.[3]

Also suggestive that we may have reached some sort of peak in farmland is the slowdown in the rate at which new land is being brought into production in the most recent fifteen-year period. In the fifteen years

Table 5 CHANGES IN POPULATION, FARMED AREA, AND FOOD INTAKE, 1990–2005

	1990	2005	Change (%)
World farmed area			
(thousand hectares)	4,858,103	4,945,770	+1.8
(thousand acres)	12,004,633	12,221,263	
Arable area			
(thousand ha)	1,404,175	1,411,117	+0.5
(thousand acres)	3,469,791	3,486,946	
World population	5,284,728	6,514,752	+23.2
Consumption			
(calories/person/day)	2,700	2,800	+3.7
Total food demand			
(billion calories/day)	14.3	18.2	+27

SOURCE: FAOSTAT 2009.

from 1975 to 1990, the world's farmed area grew by 4.7 percent, according to data from the Food and Agriculture Organization (FAO). In the following fifteen years it expanded at less than half this rate (1.8 percent). Such numbers point like an emergency flare to factors that are slamming the brakes on the expansion of the world's farmed area—and raise concern over our ability even to maintain the present stock of land in a condition fit for food production.

A further pointer to approaching global land scarcity is the rate at which certain countries are now buying up "spare" farmland in others. China is said to have bought or leased 1.24 million hectares (3 million acres) of land in the Philippines and 700,000 hectares (1.7 million acres) in Laos; its Ministry of Agriculture has proposed foreign land acquisition as an explicit strategy, similar to its acquisition of global energy resources. The United Arab Emirates has acquired 900,000 hectares in land-scarce Pakistan and 378,000 in Sudan. South Korea is also reported to have acquired 690,000 hectares in Sudan. Other countries reported to have bought up foreign land—mainly in Africa—for food or biofuel production include Saudi Arabia, Malaysia, Qatar, Bahrain, Kuwait, India, Sweden, Libya, Brazil, Russia, and the Ukraine, prompting the FAO to warn of an era of "neocolonialism." Not only countries are involved in the global land grab: in 2008 the Korean corporation Daewoo was reported to have leased 1.3 million hectares of land in Madagascar, half that country's arable area.[4]

DEGRADATION AND DESERTIFICATION

Land degradation is the loss of the ability of land to produce food, either temporarily or permanently, or to maintain its natural landscape function. It usually involves the removal of surface vegetation and the loss of organic matter (carbon), soil, and plant and animal species. As depicted in John Steinbeck's famous Dust Bowl–era novel *The Grapes of Wrath,* it brings great suffering to afflicted communities, often forcing them to abandon their farms. Today land degradation is no longer a local or even a regional issue—it is rapidly assuming the stature of a global crisis by virtue of its sheer scale.

Soil degradation is not a sexy topic, yet it has been a silent force in dragging down crop yield improvement in recent decades and is a cryptic contributor to low grain stocks, rising food prices, hunger, and global insecurity. In the 1980s and 1990s it was an issue of notable concern to many governments but, although the agencies and scientists who watch such things have continued to issue warnings, in recent times the world at large has tended to treat land degradation as a low priority. In Asia, Africa, and South America, soil losses due to erosion average 30 to 40 tonnes per year for every hectare (13–18 U.S. tons/acre)—which is thirty to forty times greater than the rate at which soil naturally forms. On slopes and in severely degraded rangelands, the losses can be as high as 100 tonnes (110 U.S. tons) of soil per year. On Gondwana continents—India, Australia, Africa, and South America—soil formation rates are close to zero, so any loss amounts to mining the resource. To the question "Why should I care?" the simple answer is that degraded land produces less food and so contributes to rising prices as well as to conflict and refugee crises, thus affecting everyone. It also damages the environmental systems that support us by providing clean water, absorbing carbon dioxide, supplying timber, and supporting wildlife and wilderness.

A full generation has passed since the last major on-the-ground check was run on the health of the world's farming and grazing lands, and its data are now long out of date, showing how an issue once high among international concerns has slipped into obscurity. At that time, the Global Assessment of Human-Induced Soil Degradation (GLASOD) study found that about 15 percent of the world's total stock of land was degraded. The most severely affected regions were, in order: Europe (25 percent of land affected), Asia (18 percent), and Africa (16 percent). The main causes were loss of soil by wind or water, loss of fertility, physical problems, salinization, and industrial pollution.[5]

More recently, the FAO has conducted a worldwide survey using satellite images—a completely different technique from that used by GLASOD—which measure "greenness" or the extent of vegetation cover to assess how much the land is capable of producing. This revealed that the degraded area had increased alarmingly during the period 1980–2003, by the end of which it accounted for almost a quarter of the world's land surface. "Land degradation is cumulative—this is the global issue. The 1991 GLASOD assessment indicated that 15 per cent of the land surface was degraded; the present assessment identifies 24 per cent as degrading but the areas hardly overlap, which means that new areas are being affected. Some areas of historical land degradation have been so severely affected that they are now stable—at stubbornly low levels of productivity," the researchers commented.[6]

"Almost one fifth of degrading land is cropland—more than 20 per cent of all cultivated areas; 23 per cent is broadleaved forest, 19 percent needle-leaved forests, 20–25 per cent rangeland," they added. The study also noted, on the more hopeful side, that 16 percent of land was displaying improved productivity, and that this was mainly in cropping areas—encouraging evidence that land degradation can be arrested and even reversed with the right sorts of farming practices.[7] As seen from space by satellite, the most severely affected areas are

- Africa south of the equator (13 percent of the world's degrading area),
- Indochina, Myanmar, Malaysia, and Indonesia (6 percent),
- South China (5 percent),
- North and Central Australia and parts of the Great Dividing Range (5 percent),
- the South American Pampas (3.5 percent), and
- high-latitude forests in North America and Siberia.[8]

The countries with the worst land degradation as a proportion of the affected world area are Russia, followed by Canada, the United States, China, and Australia. The countries with the worst losses of primary productivity are Canada, Indonesia, Brazil, China, and Australia. This is of particular significance because Canada, Russia, and Brazil are often pointed to as major potential food bowls of the future (though the degradation seen by the satellites is mainly of forests). An important finding in the study is that although farming is a major contributor to land degradation, forestry and grazing are worse.[9] Indeed, in many croplands in the third world, degradation has slowed and sometimes even reversed, whereas the loss of forests and grasslands continues to accelerate.

Overall, the research suggests, humanity is degrading about 1 percent of its productive land area every year. This may not sound like much—but, left unaddressed, it will ruin two-thirds of the world's productive land by 2050. This will obliterate gains made by expanding the area farmed and improving crop yields.

In total, the report calculated, manmade land degradation affects more than 350 million square kilometers (135 million square miles) of the Earth's surface, has a direct impact on one person in four, and has caused the loss of almost a billion tonnes (1.1 billion U.S. tons) of carbon—the prime source of fertility—from the affected soil. This last statistic flags not only the enormous loss of food potential but also another significant threat: the capacity of land degradation to drive climate change by releasing huge amounts of carbon into the atmosphere. The more we degrade the land, the faster we warm the planet. Indeed, as much as one-third of humanity's total carbon emissions may come from the way we manage (or mismanage) land, and if we ceased burning fossil fuels tomorrow this source of carbon emissions would continue to drive the greenhouse effect. This is an intractable aspect of climate change that rarely achieves prominence in the public debate, but it is caused by every one of us as we eat.

Typifying the degradation of the world's farming and grazing lands is Inner Mongolia, in the northern part of China. Here is a landscape now often totally devoid of vegetation, says the farming systems scientist David Kemp, who recounts,

> I saw it at the start of summer. . . . [T]here had been no green forage and almost no dead forage in the area for the previous nine months, yet livestock were typically taken out daily to try and find something to eat. Our work recorded that over autumn, winter and spring the animals typically lose 20–30 per cent of their bodyweight—in a normal year. Droughts and bad winters are worse. The classic quote we have heard from old-timers is "When we were young (forty or fifty years ago), we had trouble seeing the cattle in the grassland. Now we can see the mice." It is likely that these areas grew grass up to a metre in height, whereas the peak now is ten centimetres.

It is not hard to imagine how such regions can become permanent desert. Desertification in the dry subhumid, semiarid, and arid regions continues to worsen, contributing just over one-fifth of the total decline in net primary productivity observed in the ISRIC–World Soil Institute study. This may sound small but it is no cause for complacency, as these desert and semiarid areas cover two-fifths of the Earth's land area, are inhabited by two billion people, are already relatively heavily degraded, and have low or very low productivity. The UN Environment Programme has estimated that, due to desertification and drought, these areas experience an annual loss of food worth $42 billion. Also there is a threshold effect, meaning that once these drier areas are turned into a desert it is incredibly difficult, if not impossible, to restore them to a productive and fertile state again. These dry regions are expected to expand significantly under climate change while their populations are also enlarging more rapidly than those of temperate regions. As much as a third of the world's nine billion people will live in deserts by 2050.[10]

TOXIC SOIL

Soils that contain too much salt, are too acidic, or are heavily polluted with industrial wastes are generally toxic to plants, and often to the animals and people who rely on them for food. Industrial pollution, in various ways, is increasing the amount of land and water that is unfit for growing food. Like soil degradation, this is a global epidemic but one that is subtle, slow-burning, deadly—and of which nobody knows the true extent.

Salinity can be either natural or manmade and affects about a tenth of the Earth's land surface. Naturally salty areas don't produce much food, but in the dry farming regions of the world, human-caused salinity—resulting mainly from removal of deep-rooted trees and poor design of drainage or irrigation—is spreading like a lethal crystalline shroud, taking much good land out of production. According to Michael Stocking, an expert in land degradation and sustainable land management, the 1990 GLASOD study put the area affected by salt at 4 percent of the world's total land area, and 7 percent in Asia. The problem has undeniably become far worse since then, taking much valuable land out of food production, but no one really knows how much worse. Soil salinity may seem a lesser factor among the drivers of global food insecurity, but one of the hardest-hit regions is the Indus valley, breadbasket to both Pakistan and India, where it has been estimated that Pakistan alone sustains salt damage to around 40,000 hectares (about 100,000 acres) of irrigation land a year.[11] Salinity is thus a potent ingredient in the cauldron of tensions over food, land, and water between these two nuclear powers.

Acidic soil is another time bomb. Each time a crop is harvested or a pasture grazed it removes certain elements, with the result that the soil turns gradually more acidic until it eventually reaches a point where food crops will not grow. This process is hastened by the use of fertilizers, by legume rotations, by other human activities, and by acid rain. In intensive and prosperous farming systems surface soil acidity can be countered by applying lime or gypsum, but in countries where farmers cannot afford to do this—or where lime is not readily available—it becomes ever more difficult to sustain yields, even when growing acid-tolerant crops. The area of the world's acid-affected farm soil is estimated to be around two million square kilometers (three-quarters of a million square miles).[12] One of the worst manifestations of soil acidity occurs when tropical rainforests are felled, leaving soil so acid that few plants or trees can grow; after a brief year or two of farming, the area is colonized by short-lived grasses of little nutritive value and becomes what has been dubbed "the green desert." Not only is this happening throughout Asia, where tropical forests have been cleared on a massive scale, but it could also potentially affect large expanses of Latin America and Africa if savannah and jungles were to be further cleared for farming. Acidity is thus one of the main environmental obstacles to unbridled expansion of agriculture in the great tropical river basins.

The arsenic epidemic in Bangladesh, West Bengal, and other Indian states, China, Chile, Cambodia, Laos, Burma, Pakistan, Nepal, Vietnam, Taiwan, Iran, Argentina, Finland, and the United States poisons around one hundred million people every day through toxic food and drinking water.[13] This epidemic is due to the extraction of groundwater causing the sediments to dry and release soluble forms of arsenic; these then dissolve into the groundwater when the wet season recharges it. The toxic water is then used for household purposes and irrigating food crops.

Industrial contamination of soil is a growing problem worldwide—and an emerging human health issue; some scientists link the rising global epidemic of cancers and degenerative diseases to the toxic cocktail of pollutants encountered in the home and in the air, water, soil, and food of city dwellers. This consists of heavy metals in the waste streams of mineral processing and factories, spills and leaks of fuel and industrial chemicals, timber treatments, the overuse of agricultural pesticides and fertilizers, poor disposal of sewage, and industrial fallout from the air. Because cities, which produce most of these contaminants, are usually built in river valleys they have a tendency to pollute the best food-growing soil, either directly or through the movement of the toxic contaminants in groundwater and air. For example, in recent decades it has been recognized that ozone emitted by industry in large cities damages crops growing downwind, shutting down the plants' ability to turn sunlight into food energy. This can reduce yields by 10 percent or more.[14]

The GLASOD study put the area of world farmland contaminated by industrial pollution at 1 percent in 1990, though it ranged up to 8 percent in industrialized societies such as Europe. This has undoubtedly expanded dramatically in the intervening years with the heavy industrialization that has taken place in Asia and elsewhere. The *China Environmental Times* quoted China's State Environmental Protection Administration as saying that 100,000 square kilometers (39,000 square miles) of cultivated land are polluted, contaminated water has been used on a further 21,000 square kilometers (8,100 square miles), and another 1,300 square kilometers (500 square miles) have been destroyed by toxic waste. "In total," it says, "the area accounts for one-tenth of China's cultivatable land, and is mostly in economically developed areas." The report continues, "Soil pollution presents a genuine danger. An estimated 12 million tonnes of grain are contaminated by heavy metals every year, causing direct losses of 20 billion yuan (US$2.57 billion). Harmful substances accumulate in crops and, via the food chain, find their way into our bodies, where they can cause a variety of illnesses. Soil pollution also damages ecosystems and ultimately threatens their safety. Measures to prevent soil pollution are weak in China," it warns.[15] China, almost alone among the newly industrializing countries, has recognized the true implications and vast extent of this problem.

CITY SPRAWL

Most cities occupy fertile river valleys and coastal plains for the historical reason that they needed to be close to their food supplies. As these cities expanded they encroached on more and more of the land around them, drawing their food from increasingly distant places. The fact that farmers at the city edge sold their land to developers, householders, and industrialists and moved away seemed unimportant—until the twenty-first century, when the urban population overtook the rural population globally for the first time, the world began running out of land, and

Table 6 POPULATION AND LAND-USE PROJECTIONS FOR THE WORLD'S LARGEST CITIES, 2030

	Population (in millions)	Built-up area (square kilometers/square miles)
Jakarta	37	2,720/1,050
Tokyo-Yokohama	36	7,835/3,025
Manila	36	1,425/550
Mumbai	30	777/300
Delhi	30	1,425/550
New York	20	11,264/4,349

SOURCE: Demographia, "Demographia World Urban Areas: 2025 and 2030 Population Projections," tables 4 and 8, August 2008, www.demographia.com/db-worldua2015.pdf.

food security started to unravel. If you add together all the built-up areas in the world, it is estimated they would occupy 4.75 million square kilometers,[16] a virtual city half as large as the territory of the United States or the People's Republic of China—*almost entirely built on the world's best farmland.*

By 2030, each of the megacities (cities with populations greater than twenty million) will occupy a metropolitan land area between 700 and 12,000 square kilometers (300–4,500 square miles) in extent (table 6). Beyond this, they will have even larger periurban "commuter belts," which also swallow up good farmland and price it beyond the reach of food producers. In all, it is estimated, between 20,000 and 40,000 square kilometers (77–154 square miles) of good arable country is turned into "concrete jungle" every year.[17]

But cities have an even more subtle and pernicious form of sprawl, which engulfs whole landscapes: the spread of land-hungry recreational activity in the form of golf courses, horse farms, weekend retreats, resorts, and holiday villages. To map the built-up area of a city, or take a satellite image of it, gives little impression of the vast tracts of surrounding local farming land within its "catchment" that it has diverted to these pleasant but nutritionally unproductive pastimes. This area of land removed from agriculture may be from three to ten times the actual extent of the built-up urban footprint.

An additional, unmeasured, factor is that by annexing the best farmland around the urban periphery, cities drive food production out into more marginal country, thus compounding the dangers of soil degradation, increasing "food miles," and magnifying the risk to the food supply from climate variability: to maintain the same output of food, a lost hectare of fertile river valley land close to a city may need to be replaced by five or even ten hectares of semiarid, drought-prone country that is poor in nutrients, at high risk of erosion, and thousands of kilometers away.

At the same time, modern cities, which once supplied quite a lot of their own food, especially in the form of fresh fruit, vegetables, and poultry—notably in Asia—have lately been planned and developed in ways that expel agriculture from within the urban perimeter. This is a piece of extraordinary blindness on the part of today's urban planners (to which we will return), which could well turn some of these giant cities into death traps in the event of serious future disruptions to food supplies.

City sprawl is a species of global land degradation for which there is no easy solution. Farmland is cheap; cities can afford to buy it for other purposes and farmers can seldom afford to buy it back. Also, city authorities are under relentless pressure from developers, local politicians, and home-buyers to "release" more land—a term that must from now on be clearly understood as a euphemism for the destruction of its food potential. In the coming famine of the midcentury, the ability of cities to grab vitally needed farmland for nonfood uses may be viewed as a serious market failure requiring legislative correction: countries and regions concerned about their loss of food security may need to consider placing legal bans on the sale of arable land for nonagricultural purposes.

The pressures that urban spread creates are highlighted in countless clashes between farming people and authorities or land developers around the world, which invariably end with the farmers being put off their land. Perhaps the most celebrated has been the long-running battle between farmers and construction authorities at Japan's Narita airport—a minor war that began in 1966 and was still flickering four decades later.[18] A similarly impassioned resistance was staged more recently by thousands of farmers around Dongnangang village, in China's northeast, who claimed that 100,000 hectares (247,000 acres) of their land had been stolen by corrupt authorities and who posted a manifesto on the Internet announcing that they were taking it back into private ownership.[19] In such encounters the farmers almost invariably lose—but so too does world food security. Although a single acquisition of farmland by city authorities has little direct impact on global food security, developers, like a plague of mice, nibble away ceaselessly at the most productive land on the planet, destroying our food security through the tyranny of a million ill-judged decisions.

SETTING LAND ASIDE

It is sometimes argued that land set aside in the United States and Europe for conservation purposes is available to grow more food should the need arise—but since most of that land was brought back into production when grain prices started to climb in 2007–8, and since much of the land is second-rate anyway, this is not now thought to provide much of a buffer against global food shortages. Where land has been converted back to wilderness there may also be conservation objections to its reconversion to agriculture.

FOREST IMPACT

Thirty percent of the Earth's land surface is covered by forest—which is about one-third less than originally existed before we invented the ax. The Earth's total forest area today is about four billion hectares (one billion square miles), but only one-fifth of this remains relatively intact and free from exploitation. Deforestation is driven mainly by the need of poor people to clear land for farming and by the often uncontrolled harvesting of timber—and this devours around thirteen million hectares (five million square miles) per year from the global area of forest. With the advent of world awareness about the environmental perils of clearing forest and of climate change, however, there has been a marked increase in the rate of tree planting and forest regeneration in recent years. The United Nations Environment Programme estimates that this has almost halved the net rate of forest loss to 7.3 million hectares (2.8 million square miles) per year.[20]

Although this is tremendously heartening from the perspective of restoring forests, it also indicates that less new land will be converted from forest to agriculture than in the past—and that the necessity to protect the world's remaining forests will conflict sharply with the notions of those who believe that clear-felling the Amazon, the Congo, and Siberia is the answer to the food challenge. In a world in which many people, rich and poor, are struggling to put back trees, razing vast areas of forest for farming is increasingly unacceptable—in terms of the environment and its species, the climate, and the billion people who depend on forests for their livelihoods. In 2008, for example, Norway offered Brazil $1 billion not to cut down more of the Amazon forest.[21] This suggests strongly that in the twenty-first century we may have to look to other measures to obtain our food than to the ancient resort of plying the ax.

SEA-LEVEL RISE

The Intergovernmental Panel on Climate Change (IPCC) expects sea level to rise by between 20 and 60 centimeters (8–24 inches) by the end of this century due to the effects of global warming. The effect of this on low-lying islands has been widely canvassed in the media and policy circles; less so the impact on the delta regions of the world's great river systems, such as the Ganges, the Nile, the Mekong, the Irrawaddy, the Amazon, the Yellow, and so on. These deltas have some of the Earth's richest and most productive soil, flushed out of river catchments over eons. They are also extremely low-lying, in many cases a meter or less above the high-tide mark. The conservative scenario for climate change thus carries with it considerable added risk of the drowning and salinization of farmers' fields in a number of food-critical regions due to sea-level rise directly, to the expected increase in storm frequency and ferocity, to increased flooding of rivers held back by the higher sea level, and to the intrusion of saltwater underneath farmland. For example, a sea-level rise of just forty centimeters (fifteen inches) in the Bay of Bengal—the midrange in the IPCC's scenarios—would put 11 percent of its coastal land underwater, displace thirteen million climate refugees, and wipe out one-sixth of the Bangladeshi rice harvest. A fifty-centimeter (twenty-inch) sea-level rise would displace around fifty million people globally.[22]

The spillover effect of abandoning some of the world's richest food-growing regions will be felt by all, warns the Australian climate economist Ross Garnaut.

> If sea level rises and displaces from their homes a substantial proportion of the people of Bangladesh and West Bengal, and many in the great cities of Dhaka, Kolkata, Shanghai, Guangzhou, Ningbo, Bangkok, Jakarta, Manila, Ho Chi Minh City, Karachi and Mumbai, it will not be a problem for Bangladesh, India, Pakistan, China, Thailand, Indonesia, the Philippines and Vietnam alone. If changes in monsoon patterns and the flows of the great rivers from the Tibetan plateau disrupt agriculture among the immense concentrations of people that have grown around the reliability of water flows since the beginning of civilisation, it will not just be a problem for the people of India, Bangladesh, Pakistan, Vietnam, Myanmar and China. The problems of unmitigated climate change will be for all humanity.[23]

This is nothing, however, compared to the scenario of acute sea-level rise unleashed by the melting of the Greenland ice cap along with significant parts of the polar ice such as western Antarctica. If Greenland's

2.85 million cubic kilometers of meltwater were added to the world's oceans, that would raise sea level by 6.5 meters (21.3 feet), which would flood vast areas of farmland on coastal plains and estuaries, with catastrophic consequences for food production.[24] Indeed many scientists now consider the IPCC predictions far too conservative, based on observed sea-level rise that is toward the extreme end of expectations. If this continues, they say, an overall rise of the order of five to six meters by the end of the twenty-first century is possible. Sea-level rises of this magnitude, however, may take years or decades to develop and so provide plenty of warning. To balance these losses of farmland, melting tundra will also, over a century or so, open up fresh grain and grazing lands—though this will probably not occur quickly enough to feed humanity through the 2050 peak in population and demand.

THE LAND MYTH

It is sometimes asserted in agricultural policy reports that the world has plenty of spare farmland. Such views have bred a dangerous complacency that is partly responsible for the food crisis of the first decade of the twenty-first century, as these views conveniently ignore all the difficulties, risks, and environmental consequences—including accelerated climate change—of removing forest or grassland from vast new regions.

Brazil, for example, has the Cerrado, an area of 207 million hectares (800,000 square miles) of tropical savannah in the southwest of the country with regular rainfall, which is already delivering promising crops of soybeans, maize, and other grains. Brazilian scientists estimate that the Cerrado could yield around 350 million tonnes of grain per year—which sounds like a prodigious quantity but, seen in the context of world demand, would provide feed for only one-sixth of the livestock required to meet the forecasted growth in meat demand, and would not even replace the grain soon to be burned as biofuel. Also, the Cerrado's soils are acidic, requiring heavy applications of lime, and low in nutrients, requiring heavy fertilization—meaning that the Cerrado is both high-cost and high-risk to develop. The Australian agricultural scientist Tony Fischer comments, "The Cerrado region will make an important and growing contribution to the world's food and feed supplies, but not the huge contribution that we will need all too soon."[25]

The European Bank for Reconstruction and Development considers that there are 23 million hectares (88,000 square miles) of good farming country lying idle in Russia, Ukraine, and Kazakhstan, of which just over half could be brought back into production fairly quickly and safely, boosting the harvest by at least 30 million tonnes (33 million U.S. tons) and making Russia the world's second largest grain exporter.[26] Even combined with Brazil, however, the total potential of these "food bowl" regions falls well short of the two to three billion tonnes of additional grain needed to feed humanity in the midcentury—and other measures will be needed, particularly in the form of higher yields (see chapter 7).

SOLVING PEAK LAND

Peak land is an issue that some experts seem eager to downplay or dismiss because their maps appear to suggest that the Earth has plenty of spare terrain. As we have seen, however, there are serious impediments, costs, or risks to opening up much of this land and, if the experience of recent years is anything to go by, it will be opened only at rates far too slow to match the rising demand for food on its own.

This, essentially, forces us back to four main strategies:

1. Redouble output from existing land and water using better and more sustainable farming methods, both small-scale and large (see chapter 7).
2. Make a global effort to turn lost or degraded land back to productive use.
3. Develop new ways to produce food that don't take up a lot of room.
4. Stop wasting food.

Some people would add: reduce the population. Although this is very important to addressing the global food challenge (see chapter 10) in the longer term, this is unlikely to happen in the next half century, short of a catastrophe.

Lift Output

The main ingredients in lifting farm output per area of land are science and technology, skillful farmers, and careful management of plant nutrients. Since 1961 this combination has been largely responsible for boosting global food output by 178 percent and crop yields—how much we grow on a given area of land—by 143 percent.[27] Because this formula has worked in the past we should undoubtedly keep on using it, with even greater care for the environment, for rural people, and for the sustainability of food production into the future. Yet the world and its governments have retreated from this winning strategy: the 2007–8 food crisis was the warning bell. Also it is clear that this form of energy-intensive farming feeds only part of the world's population, even if it has the capacity to feed all, and, as we will see in later chapters, it depends critically on two essential but finite resources—fossil fuel and mined nutrients—that will inevitably be exhausted, requiring us to find different answers.

To feed the projected human population in 2050 requires us to lift total farm output by around 2 percent a year. This may not sound like much, but it has to be achieved in both rich and poor countries, by all sorts of farmers in all sorts of environments, and in the face of growing scarcities of land, water, and (as we will see) nutrients and energy. Presently the world is achieving only about half the increase necessary.

Repair Damaged Land

Of central importance is the development and dissemination of farming techniques that do not deplete or damage the soil or pollute or waste water, that maximize food output with the fewest inputs, and that have sharply reduced impact on the wider environment. This is known as conservation farming and it is being adopted, at varying rates, all around the world. Conservation farming means disturbing the soil as little as possible in order to plant a crop and control weeds; keeping a cover of crop, pasture, or stubble on the ground to build organic matter and prevent erosion; preserving soil moisture; and rotating among different crops so the soil does not become depleted in key nutrients and preserves a rich diversity of life (such as worms and beneficial microbes). It does not necessarily involve heavy use of fossil fuels, chemical pesticides, or manufactured fertilizers, although these form the basis of most high-production farming systems and underpin global food security. Sharp spikes in world oil and fertilizer prices between 2006 and 2008 highlighted the need to design low-input farming systems for both large- and small-scale agriculture. At all scales, conservation farming relies for success on the development of high-yielding, easy-care crop varieties that are capable of withstanding drought, insect attack, and disease; are easy to store or handle; and are more nutritious to eat. It relies on the ability to reincorporate crop wastes back into the soil to make it more fertile. Conservation farming is also important as a means of locking up more carbon from the atmosphere in the soil, making efficient food production a potential way to counter global warming.

In the rangelands, where much of the world's livestock are produced, conservation farming means matching the numbers of animals grazed to the long-term ability of the natural vegetation or pasture to recover and support them, while maintaining the health of the whole ecosystem, including its wildlife. Only lately have tools been developed to give pastoralists a clear understanding of what their land can carry under an erratic climate, when to breed up and when to sell off surplus stock. At present most of the world's rangelands are being flogged, as the instinctive response of most graziers and herders to low incomes is to carry more animals. Research in China and elsewhere, however, has shown that by reducing sheep or goat numbers, graziers can increase both meat quality and their incomes while restoring the grasslands to a more productive state.[28]

Recently an unfortunate ideological rift has opened up between adherents of "organic" or small-scale, low-input farming and large-scale, high-input farming. The truth is that the world will need both schools of agricultural thought in order to feed us through the midcentury peak in food demand. It will need high-input agriculture in order to grow far greater volumes of food from the limited land and water available to support the teeming urban populations—and it will need science-based organic and low-input systems because these are what most of the world's farmers use and can be highly productive; they are also sustainable because they mainly use human labor instead of fossil fuel–based inputs. The ideological divide between high-intensity and smallholder farming is epitomized in the fierce public and media debates that have taken place in many countries over genetically modified crops, the use of pesticides, and the role of agribusiness. Regardless of the vehemence with which either side argues its case, it is highly unlikely the world will come down in favor of one and reject the other. To forcibly return the world to a condition of small-scale, low-input farming would be a prescription for mass starvation. And to turn global agriculture over wholesale to intensive modern broadacre farming would throw a billion subsistence farmers off their land, as well as expose the entire food supply to shortages of nutrients or fossil fuels.

Proponents of both forms of agriculture need to join forces to solve a common challenge facing the human race rather than decrying each other's approach: organic, low-input, and smallholder farming systems need to become more science-based in order to raise yields reliably and lose less food post-harvest—and advanced farming systems must seek ways to produce more food with far less energy, water, land, and chemicals. Above all, both schools must continue to search ceaselessly for ways to minimize their impact on the wider environment and to share the resulting knowledge freely with each other.

That smallholders have a place in global agriculture is clear from evidence that the smaller the farm, the more food it tends to produce per unit of land.[29] This has sparked a fifty-year dispute between advocates of large-scale and small-scale agriculture on which, to be frank, the jury is still out so far as feeding the world is concerned. It may well turn out that productive smallholders are needed to feed their families and local communities, whereas high-intensity broadacre farming and agri-food systems need to lift productivity to feed the world's burgeoning urban populace. Each, at present, feeds about half the world's people, and both are essential to our future. Arguing about which is the better of the two systems and should thus become the global model for food production is therefore not only pointless but, in the present context of emerging shortages, risky. Both need far more support—and more than lip service from governments.

A critical issue emerging from this debate is that security of land tenure is of tremendous importance in ensuring the food supply, and farmers—whether large or small—find it much harder to be productive if they do not have control over the land they farm. Also, farmers with no land tenure seriously consider why they should spend money and "sweat equity" to care for the land and improve its quality, if it can be wrested from them by a third party. Indeed, improved land is more likely to be seized than unimproved land—so hard work means that good tenant farmers are more likely to lose their land. This is an issue so deep and wide, and which varies so greatly from country to country, that it cannot be tackled in a book of this nature but must be solved by each individual nation and farming community in its own way. It should be clear, however, that whether wealthy landlords (as in large parts of the third world), centrally planned governments (such as North Korea's or Burma's), or interfering bureaucracies (found just about everywhere) are the cause of fragile land tenure and loss of freedom to farm efficiently, they are all in their different ways contributing to global food insecurity. Liberated of these strictures, smallholders and large-scale farmers are both capable of producing a lot more food.

A promising development is the emergence of a worldwide permaculture movement, based on the ideas of two Australians, Bill Mollison and

Ian Holmgren. Permaculture (or permanent agriculture) is an approach to designing human settlements, in particular perennial food-, timber-, energy-, and fiber-producing systems, that mimic the ecological interrelationships seen in nature. Although its systems tend to be micro in scale and to suit best those who wish to be self-sufficient in the production of their own food, energy, and other living materials, permaculture embodies many of the principles that much larger farms, farm landscapes, catchments, and entire regions need to embrace if they are to manage their resources well into the future.[30] As a system, it merits greater scientific study and support and is an example of how small can indeed be beautiful—but has the potential to make big beautiful as well.

Two other solutions to the constraints posed by limited farmland, which will be discussed later, are the greening of cities in order to make them more self-sufficient in food production, improve their amenity values, and end their colossal waste of water and nutrients, and the development of intensive biocultures producing nutritious food from what is presently regarded as waste.

Peak land, like peak water, is a wake-up call that warns us that we are approaching the limits of the Earth's ability to support us. Though rarely mentioned in the public debate, the scarcity of farmland is likely to emerge as a fresh source of international tension and possible conflict as the century advances.

WHAT CAN I DO ABOUT IT?

Although many of the issues discussed in this chapter relate to national and global challenges and solutions, there are nevertheless things individuals can do in their own lives to ease the pressure on the world's farmland. These include:

1. Avoid wasting food.
2. Eat more grains, fruit, and vegetables because these take less land to produce than do meat and dairy.
3. Support politicians and governments who are prepared to put more scientific effort into farming, to resolve land-tenure injustices, and to reduce overregulation of food production.
4. Be selective as a consumer in choosing those foods that use less land, energy, and water to grow. Demand eco-labeling from supermarkets and manufacturers.

5. Support green cities to reduce the city footprint on the wider landscape and reuse its wastes.
6. Support the rezoning of periurban landscapes to preserve agricultural values and curb land-hungry development: it will keep your own food prices down and help prevent desertification.
7. Grow more of your own food as efficiently as you can, composting safe household waste.

FIVE

NUTRIENTS—THE NEW OIL

> There are no substitutes for phosphorus in agriculture.
>
> **—U.S. Department of Agriculture**

Britain has declared war on waste. In 2008, Prime Minister Gordon Brown urged his fellow Britons to stop throwing away their food, as a measure to head off the global food shortage. He went on to put the case to leaders of the world's economic superpowers at the G8 summit in Japan, where they listened politely as they downed their sumptuous banquet.[1]

Ours is the most profligate generation in history. We waste food and, even more important, nutrients as if they were infinite and inexhaustible. As if there were no hungry people in the world and as if there were no coming famine. We act toward food and the land in ways that would utterly appall and horrify our ancestors, accustomed as they were to husbanding and recycling every precious form of "waste" in order to turn it back into food.

Brown's anxiety was well-founded. A former government food advisor, Lord Haskins of Skidby, who worked for one of the nation's largest food suppliers, had calculated that 60 million Britons were each year wasting around 20 million tonnes (22 million U.S. tons) of food—16 million tonnes in homes, shops, supermarkets, wholesalers, markets, and manufacturing establishments, and around 4 million tonnes on the farm or in transit. The average household could save $1,000 a year on food purchases if even a fifth of this wastage could be eliminated. The chief culprit, it turned out, was the use-by date, which was causing consumers to throw out one-third of all the food they bought.[2]

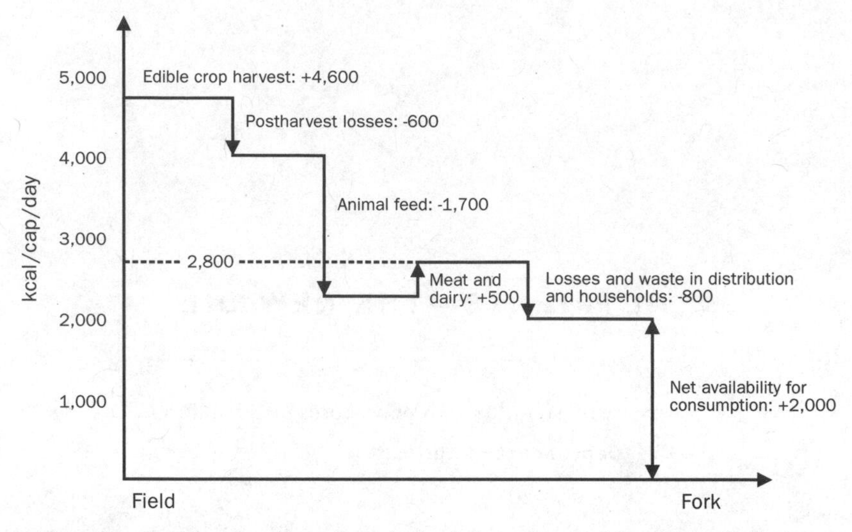

Figure 1 Food produced and lost along the food chain. Source: J. Lundqvist, C. de Fraiture, and D. Molden, "Saving Water: From Field to Fork—Curbing Losses and Wastage in the Food Chain," Stockholm International Water Institute Policy Brief, 2008, fig. 1, www.siwi.org/sa/node.asp?node=343.

The squandering of food is by no means confined to Britain. In the United States, Timothy Jones of the University of Arizona at Tucson has found that America wastes 40–50 percent of all its food. The typical American household was trashing 14 percent of its food purchases, worth $590 a year, including products whose use-by date had not yet expired.[3]

That rich countries and consumers waste food is scarcely news, though the prodigious volumes involved are only now starting to impinge on the public consciousness—and then only in a handful of countries. Waste of food is not confined to the affluent world, however: globally, the UN Food and Agriculture Organization (FAO) estimates, a full third of all fruits and vegetables never reach the consumer at all, perishing in the fields, in storage, or en route, and being attacked by a ravenous horde of molds, insects, and rodents. "Spoilage claims 30% of India's fresh produce, while postharvest losses of fruit and vegetables in some African countries can reach 50%," the World Vegetable Center notes.[4]

The Stockholm International Water Institute (SIWI) goes further, estimating that more than half the world's food is wasted in one way or another. We harvest around 4,600 calories per person per day, it says—but only 2,000 of these actually reach our mouths, as figure 1 shows.

In SIWI's report "Saving Water: From Field to Fork," Jan Lundqvist and colleagues argue that a loss of food is equivalent to a loss of water—and if we want to stop wasting water, one practical way to do so is to stop wasting food. They say that the goal of reducing food waste by half is both necessary and achievable. This argument applies with equal force to nutrients and fertilizers.[5]

There are tragic ironies here: a world in which a billion people go to bed hungry throws away food sufficient to feed three billion. A species that has used its brains to treble food output in two generations is chucking half its achievements into the garbage can. And if there were no waste, there would be no food insecurity and a far lower risk of war, famine, or refugee crises. This is a case where small changes in individual, corporate, and regulatory behavior and government policy can literally change the world for the better.

If it were easy to stop wasting food, however, we would perhaps have done it long ago. Often, too, the wastage is driven by other valid concerns, such as protecting consumers from spoilage and food poisoning—and current regulations need to be revised or replaced with ones that are equally safe but far less wasteful.

FERTILIZER MADNESS

Shockingly, the immense global waste of food is but a fraction of an even more colossal squandering of nutrients, another of the principal drivers of the coming famine.

Simple nutrients support all life on Earth. From microbe to plant to human being, every living organism relies on the molecule ATP (adenosine triphosphate) as the essential energy carrier in its cells. This makes the element phosphorus indispensable to all living things.

Fertilizers made from essential nutrients such as phosphorus have powered the farming miracle that has tripled world food production in the past half century. Their use on high-yielding crops developed in the Green Revolution prevented starvation in countries such as China and India and doubled harvests in many countries. Just as oil has been the fuel of the transportation revolution, fertilizers are the fuel of the global food miracle.

Yet humanity today hemorrhages nutrients at every link in the chain. They bleed from the farm itself in soil, water, and wind. They are lost when food perishes in transit or storage. They are sacrificed when grains, fruits, or livestock are processed into food and inedible "waste" is discarded. They are squandered all along the food chain from factory to supermarket to home. They are lost in cooking. They go in our garbage bins. And when we dispose of our sewage, the nutrients it contains often go out to sea, to fertilize the deep oceans or to pollute rivers, estuaries, and coastal waters, causing choking blooms of algae, which in turn smother, starve, or poison fish and edible sea life along our fertile continental shelves.

LEAKY FARMS

Despite our prodigious waste of food, it is probable that the greatest loss of nutrients happens on the farm.

Though they garner fewer media headlines, fertilizers have shaped recent human history as much as, or even more than, antibiotics or bullets. It has been estimated that more than two billion people would not be alive today were it not for the invention of the industrial process for making nitrogen fertilizer. Indeed, fertilizers have been branded the principle cause of the human population explosion. The world's main food crops are estimated to take up around 12 million tonnes (13.2 million U.S. tons) of phosphorus every year, whereas only 4 million tonnes of phosphorus are generated from natural weathering of rock or atmospheric deposition: this highlights civilization's critical dependency on the supply of artificial fertilizers, and our increasing vulnerability to any shortfall or disruption in supply.[6]

Worldwide, farmers today use seven times more fertilizer than they did a half century ago, and food output has risen two-and-a-half fold. Fertilizers are a vital component of modern mass food-production systems in advanced countries and also a pillar of the most successful outcomes of the Green Revolution in countries such as India and China. Total world fertilizer use in 2010–11 was projected by the FAO to be 211 million tonnes (232 million U.S. tons), leaving a 19 million tonne (21 million U.S. ton) surplus (table 7).

Products derived from nitrogen (N), phosphorus (P), and potassium (K) make up nine-tenths of total global fertilizer consumption. In food production there is no substitute for these three nutrients: they are as essential to plant growth as water or light. If one of them is lacking, plants grow poorly or not at all. A long-running experiment in Kenya shows that the yield of food from a field of maize and beans is four times higher with fertilizer, manure, and stubble retention than without these ways of adding nutrients. All told, researchers have calculated, fertilizers

Table 7 WORLD FERTILIZER SUPPLY AND DEMAND (IN TONNES)

Years	Supply	Demand	Surplus
2007–8	206,431	197,004	9,427
2008–9	212,225	201,482	10,743
2009–10	219,930	205,947	13,983
2010–11	230,334	211,230	19,104
2011–12	240,711	216,019	24,692

SOURCE: Food and Agriculture Organization of the United Nations, "Current World Fertilizer Trends and Outlook 2011/12," 2008, table 2, ftp://ftp.fao.org/agl/agll/docs/cwfto11.pdf.
NOTE: 1 tonne = 1.1 U.S. tons

spared us the necessity of cutting down or plowing up an extra 2–3 million square kilometers (0.7–1.2 million square miles) of forest and grassland to feed the world during the last three decades of the twentieth century.[7]

Almost all the world's N fertilizer is made from synthetic ammonia produced using natural gas and is manufactured in more than sixty countries. The lion's share of phosphate production, on the other hand, comes from China (37 percent), Morocco and the Western Sahara (32 percent), South Africa (8 percent), and the United States (7 percent). Potash is obtained by mining potassium salts and comes chiefly from four countries—Canada (53 percent), Russia (22 percent), Belarus (9 percent), and Germany (9 percent).[8]

Besides the profound benefits they have brought to humanity, fertilizers also have a number of drawbacks: they leak into the environment, polluting and sometimes killing bodies of water; they accelerate global warming; their overuse can acidify farm soil and may have adverse human health effects; and they require large quantities of expensive and finite fossil fuels to manufacture.

Although it is difficult to make a global estimate, a number of scientific studies indicate that roughly half of all fertilizers used are not taken up by the target crop or pasture to which they were applied.[9] When you put animals into the system, the nutrient loss increases further. This suggests that potentially as much as half the world's fertilizer—more than 100 million tonnes of nutrients—may be going to waste and, worse yet, causing environmental pollution and climate warming. Another way to look at this is that we can grow substantially more food than we do today provided we can somehow capture and reuse these lost nutrients.

This is vital, because the flip side to the story about the successful use of fertilizers to raise food output in many countries is that the world is mining its scarce natural soil nutrients at an alarming rate—and has been doing so for centuries, even for thousands of years in some regions. When land is overgrazed or overcropped and erosion results, precious nutrients and fertility (soil carbon) are stripped as the soil particles blow or wash away. Typically, soil cleared for agriculture or grazing loses up to one hundred or two hundred times the amount of nutrients shed by natural, uncleared land.[10] Sometimes these lost nutrients end up on other farms, but eventually most of them are carried out to sea and sink into the deep ocean. In poor countries, which can ill afford fertilizer, there is a constant mining of nutrients from the soil, both in food and through degradation. In India's soil, for example, scientists report a widening gap between the amount of nutrients applied as fertilizer and manure and the amount removed in food and lost soil. They warn of a "serious soil quality hazard" requiring "urgent attention," adding that the losses could more than double due to increased demand for food.[11] This depletion of essential nutrients is taking place over the greater part of the world's farm and grazing lands at rates far faster than lost nutrients are being replaced, erecting a massive but largely invisible barrier to our future ability to feed ourselves.

Worldwide, scientists estimate, humanity is pumping about 150 million tonnes more nitrogen and 9 million tonnes more phosphorus into the Earth's biosphere than would occur naturally. This planetary-scale pollution is now on such a scale that it may "significantly perturb the global cycles of these two important elements," potentially leading to loss of oxygen in the oceans and mass extinction, warn Johan Rockström and his colleagues. A team from Ohio State University and the U.S. Department of Agriculture has calculated that around 20 million tonnes of nutrients are lost each year from fields growing the world's four top crops—wheat, rice, maize, and barley—alone. Though they made no estimate for agriculture as a whole, the loss is probably two to four times as great as this when other crops, grazing lands, and horticulture are factored in. The Ohio team found that "there was no country without any nutrient problems in any crop production systems in the year 2000," although the majority of the losses were in developing countries and in rice farming. Furthermore, they warned of a growing imbalance in the way the world is using fertilizers, applying far too much nitrogen and insufficient potash and other elements: in this situation, they warn, "N . . . is simply used as a shovel to mine the soil of other nutrients." They conclude, "Severe nutrient deficits of N, P, and K occurred widely in harvested areas in both developing and least developed countries, particularly in the rice and wheat production systems in Asia, Central and South America, and Africa. Continuous depletion of soil N, P, and K . . . poses a real threat to agricultural sustainability and food security."[12]

Humanity is thus trapped between the hammer of nutrient depletion from unfertilized or poorly fertilized lands and the anvil of vast losses from high-fertilizer-use systems. Add to this the waste of up to half of the food we actually produce and the loss of nutrients in our sewage waste streams, and the true scale of the global hemorrhage starts to emerge. In terms of nutrients and their ability to support us, civilization is bleeding to death.

These considerations must therefore be balanced against the FAO's reassuring assessment that there is "ample" supply of nutrients for the future. Ample if we stanch the losses and reclaim the waste, maybe. Precarious indeed should we fail to do so.

Ninety-seven percent of the world's nitrogen fertilizers are made using the hydrogen from natural gas. Availability of industrial N fertilizer is thus limited by the extent of natural gas reserves and their price; moreover, the International Energy Agency has warned that global oil and gas will peak within the decade 2010–20.[13] From this point on, industrially produced nitrogenous fertilizers will become increasingly scarce and expensive, posing a threat to crop yields worldwide unless alternatives are quickly developed.

Phosphorus and potassium fertilizers, by comparison, are both mined from rock, and although geologists say the reserves are plentiful, they are nevertheless finite and will eventually run out. Rabobank notes that the world has an estimated eighty-one years' supply of rock phosphate, provided that demand stabilizes in 2020—which it is highly unlikely to do.[14] Furthermore, the quality of the reserves will decline rapidly as miners exhaust all the high-grade material, and this will drive up costs significantly as well as consume far more fossil energy. Supplies of phosphorus are likely to become critical sooner than those of potash. All fertilizers are traded on world markets but, with just three countries controlling the lion's share of the world's rock phosphate, supply is exposed to the same vagaries as oil—wild price speculation, regional shortages, trade bans, wars, and so on—making both price and security of supply increasingly uncertain as the century advances. Also, countries with large farm sectors of their own to support, such as Brazil, China, Canada, and Russia, may in the future feel justified in reserving domestic resources for their own farmers.

Of humanity's rising dependence on artificial fertilizers, the FAO has stated, "The relative share of mineral fertilizers considering all sources of global nutrient inputs available for crop production, however, is projected to increase from 43% in 1960 to 84% in 2015."[15] This implies that synthetic fertilizers (in the absence of other nutrient sources) will be asked to shoulder the primary burden of meeting our food needs by midcentury—at precisely the time when supplies start to run low. This will render humanity, and individual nations, all the more vulnerable to anyone who wishes to hold a nutrient gun to their heads by denying supply, or to shortages caused by whatever reason. Governments may be able to withstand a grumbling citizenry in the face of shortages of oil. How long can any of them survive an angry citizenry that goes unfed? How easy will it become to undermine an enemy state just by cutting off its phosphorus?

Fertilizers are subject to rapid and unexpected price increases. During the global food price panic of 2007–8, when consumers were facing increases of a few percent in food prices, farmers were being asked to pay 160 percent more for urea (a nitrogen fertilizer), while diammonium phosphate prices skyrocketed by 318 percent. This caused many farmers worldwide to cut back on or abandon the use of fertilizer, which in turn reduced food production.[16] At the same time, high fertilizer prices also encourage the opening of new mines—which add to the global supply of fertilizer in the short run but deplete reserves more quickly in the long run.

PEAK PHOSPHORUS

A daunting question is: could the world's fertilizer supplies ever run out? In 2007 the Canadian physicist Patrick Déry decided to apply the famous "peak oil" theorem of M. King Hubbert to rock phosphate (figure 2). To his dismay, he found that the world had actually passed peak phosphate back in 1989! Until now, humanity has been comforted by vague predictions that there is a fifty to one hundred years' supply of P, and we have been negligently content to bequeath the worrying to our grandchildren. But the point about peak theory is that the trouble begins well before the supply is exhausted—in fact, it starts at the point when demand begins to outpace supply, which is where prices start to jag about, reflecting people's apprehension over approaching scarcity.

Scarier still is the fact that, although there may be energy substitutes for oil or gas if supplies run low or become too expensive, there are no

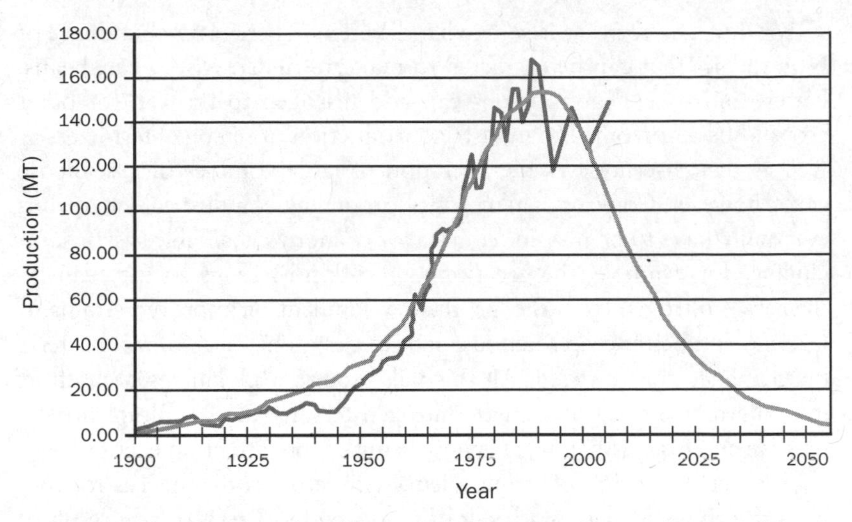

Figure 2 World rock phosphate production, 1900–2050. Source: Patrick Déry and Bart Anderson, "Peak Phosphorus," *Energy Bulletin,* August 13, 2007, www.energybulletin.net/node/33164.

substitutes for phosphorus. It is fundamental to the chemistry that supports all forms of life. It is essential to crop and pasture growth.

No phosphate, no food.

When plants do not have enough P to grow, scientists refer to them as "phosphorus-limited." You can throw all the other nutrients you like at them and they will not grow. Unless humanity can find twice as much P to sustain a doubling in food output or, better, recycle the sources we already have, the blunt truth is that human civilization as we know it is phosphorus-limited. Any biologist can tell you that this means "headed for trouble."

The FAO suggests that there is no need to panic just yet—there remains a market surplus of nutrients, including P. But such observations breed complacency: already demand is outrunning the discovery of new resources. And as food demand rises inexorably as we approach midcentury and as farming systems modernize and intensify, the quality of rock phosphate reserves will decline as the best ones are mined out, driving up prices to the farmer and escalating international tensions over what's left. People talk about the wars over oil, diamonds, or water; that nations could go to war over phosphate or potash so they can feed their children has not yet surfaced on the global agenda. It will.

For nitrogen the case is somewhat different. There is an abundance of N in the air that can be extracted to make fertilizer. Also, peas, beans, lentils, and other legume crops can add nitrogen to the soil for other crops to benefit from. The industrial production of nitrogen fertilizers as well as their use involves the liberation of large volumes of greenhouse gases, however. Conventional means of producing N industrially may thus eventually have to be phased out in favor of more sustainable systems, including, for example, the genetic modification of crops so they can fix their own nitrogen from the air, the development of improved strains of legumes and better crop rotations, and the design of suites of soil bacteria that enhance soil nitrogen. All this will require a global research effort into alternative ways to increase nitrogen fertility in agriculture no less massive and urgent than what is now being devoted to clean energy.

Another dimension of the problem, with lethal consequences for millions of people, is that, as a result of not using fertilizers, or as a result of using them badly, deficiencies in essential micronutrients are showing up. Joachim von Braun, of the International Food Policy Research Institute, says: "Micronutrient deficiencies . . . pose a vast global health problem. Vitamin A deficiency, iron deficiency anemia, and zinc deficiency increase the probability of early death for children and women, impair IQ development in children, and lead to a large loss in quality of life, productivity, and economic growth in developing countries. Vitamin A deficiency . . . leads to approximately 1 million child deaths every year."[17]

SOLVING THE NUTRIENT CRISIS

There is no answer to the challenges outlined in this chapter other than for humanity to wake up to the fact that it confronts a global nutrient crisis at least as serious for its future as the much more widely discussed issues of water scarcity, peak oil, and climate change. The greatest obstacles to this awakening at the moment are public lack of awareness and political apathy over the gravity of the situation.

A good start would be for every nation to commit to a plan for nutrient conservation and recycling. This might involve measures such as

- ending or greatly reducing all forms of soil erosion,
- recycling nutrients within the farming system on a substantial scale,
- eliminating fertilizer subsidies, which promote wasteful use, and introducing incentives to conserve nutrients,

- launching national campaigns to sharply reduce the waste of food throughout the food chain from farm to consumer,
- designing dietary campaigns to reduce the proportion of meat and dairy in the diet,
- funding a massive research effort on ways to conserve, recycle, and reuse nutrients all along the food chain, on a scale similar to the research effort now being put into clean energy,
- implementing plans to compost all organic urban waste and put it back into the food cycle,
- developing improved technologies to harvest nutrients from waste streams,
- harvesting urban sewage sludge and incorporating it (suitably purified) into fertilizers for use on farms or in urban horticulture or biocultures,
- separating urine, which is high in both N and P, from other wastes for conversion to fertilizer, and
- replacing water-based toilets with composting designs as part of a nutrient recycling system.

Such measures must then be built into a global agreement akin to the Kyoto Protocol on global climate change or the Montreal Protocol on ozone depletion.

GREEN CITIES

An alternative way to view today's megacities is as vast collecting points for water and nutrients. In order to sustain their inhabitants, cities gather water and harvest nutrients locally, across their continent, and all around the world. Then, having concentrated and consumed both nutrients and water, they mostly throw them away, often horribly polluted with toxins.

Having half the global population, cities already concentrate more than half the world's food nutrients. And within a generation they will also concentrate half the available freshwater, leaving insufficient water to spare for the farmers who are being asked to double food production. City planners and administrators, however, are for the most part not interested in capturing these precious resources in order to sustain all citizens in the future, seeing them only as "waste" to be disposed of. Yet, if the world is short of water and nutrients, it makes sense to harvest them where they are most readily and cheaply available. Within a decade or so, our cities can be turned into mines for water and nutrients. All these precious substances can then be reused again and again. Humanity is thought to produce around 3 billion tonnes (3.3 billion U.S. tons) of phosphorus in its sewage,[18] so, in theory at least, the world's cities concentrate around 1.5 billion tonnes (1.7 billion tons)—an immense resource that is largely wasted by flushing it into the oceans.

There is an urgency about this that has not so far received due attention. Half the world's people now live (and, by midcentury, three-quarters of the world's people will live) in cities in circumstances where they will be totally without the means to feed themselves. Today's megacities are founded on the complacent assumption by urban planners that, whatever happens, there will always be a vast river of food flowing in from outside every day of the year. Yet, as we have seen, this assumption rests on increasingly tenuous supplies of water, land, and nutrients and on other deepening uncertainties such as transportation fuels and world peace. Never has urban civilization as a whole been so at risk of catastrophe—yet so astonishingly blind to it. If food supplies to a major city of twenty to thirty million people were cut off even for a week or two, the consequences would be horrendous.[19]

For reasons of health and amenity, city planners, investors, and health authorities have sought to banish almost all forms of food production from within the urban perimeter, driving them farther away, into less reliable regions and even into other countries. This is a process that must now be reversed: cities must again become farms. They must recover a degree of the self-sufficiency they enjoyed in earlier phases of civilization. And they must cease squandering freshwater and nutrients as if there were no tomorrow. It is time to recognize that regulations about waste disposal and urban farming, created with perfectly sound public health aims in view, threaten far more lives in the event of major food scarcities than they will save.

There is a growing movement, led by a handful of architects and horticulture specialists, to "green" our cities. This refers less to aesthetics or the introduction of clean air, novel transportation systems, and "eco-building codes" than to the essentials of human survival: food and water. It proposes growing food crops, especially vegetables and fruits, on, in, around, and even underneath buildings.[20] It proposes recapturing water and nutrients and returning them to the wider farming landscape. It proposes developing entirely novel systems of food production, such as edible plant cell cultures, microbial protein, and artificial photosynthesis for turning hydrocarbon waste into carbohydrate nutrition.

As we will discuss later, vegetables are one of the potential solutions to the global food crisis (not enforced vegetarianism—just more veggies in the diet). The reason is that they are very efficient converters of water, nutrients, and energy into food—and they take up far less space than other forms of food. So vegetables, vines, and fruits are ideal for the urban food-production systems of the future—grown on the rooftops of factories and large buildings, down their walls, around buildings, on balconies, in parks, along highways. If there were no other reason to do it, the fact that you can save up to 10–15 percent on your electricity bill and proportionately reduce greenhouse emissions by cloaking your building in vegetation seems a good argument. But how much more attractive to live in a *real* urban jungle, alive with greenery and bird and insect life, than a sterile, windy, concrete one?

How this might be done is illustrated in part by a Japanese civil servant, Makoto Murase, who has campaigned for years for "Skywater"—the principal of catching and storing urban rainfall and stormwater in large tanks underneath buildings, in the process setting a new design standard for the city of Tokyo. This water is mostly used for toilets and gardens—but with suitable treatment it could easily be used to grow food. Urban rainfall, storm runoff, and decontaminated wastewater can all potentially be used in this fashion.

The farsighted Mr. Murase says, "Without using urban rainwater, there is no way to support the people without destroying the rural environments that provide the food." For cities with suitable aquifers beneath them, such as Adelaide in Australia, this water can be injected below ground, creating large subterranean reservoirs where a natural cleansing process occurs, and then brought to the surface again as needed.[21] The vast quantity of water used to dispose of urban waste should also be reclaimed, along with its precious nitrogen and phosphorus, and used to irrigate urban and periurban food crops. This will necessitate a ban on the disposal of toxic industrial and domestic substances into the water stream—which many cities are already starting to address—and use of advanced techniques for removing heavy metals or other toxins from human and other organic waste. Then all these urban organic wastes, including sewage sludge, can be collected and composted, turned back into fertilizer for farms and for urban food production, along the lines espoused by permaculture advocates.

By such means the "green city" of the future may become a quarter, even a third or more, self-sufficient in food production, reducing the vulnerability of its population to food insecurity and the competitive pressure it places on farmers for dwindling resources, and greatly enhancing its living environment. To do this, however, these cities should aim to capture and recycle every drop of water and every gram of nutrient—many, many times. Just as it must emit no carbon, the future city should seek to emit no surplus water, phosphorus, or other essential nutrients but rather to "close the loop" and reuse them all.

The nutrients obtained from composted city organic waste or sewage sludge can, after suitable detoxification, be returned to the agricultural landscape by whatever means are most economical—as manure and compost, or as ingredients in manufactured fertilizers.[22] This is essential if we are to arrest the centuries-long mining of our soil and reduce urban vulnerability. Composting techniques have been widely researched in both developing and developed countries but have not so far been adopted on a very large scale by cities to deal with their vast quantities of food waste. U.S. Department of Agriculture research confirms that recycled food waste has a valuable role to play in improving soil condition and fertility for future crop production.[23]

Algae farms are another important development. These can convert waste nutrients and water into food, animal feed, fertilizer, transportation fuel, and even pharmaceuticals and fine chemicals using a range of processes, but driven principally by sunlight. Algae already foul many of our rivers, lakes, estuaries, and coastline waters due to the nutrient hemorrhage we have created, so the option of deliberately cultivating the beneficial ones and thus harvesting lost nutrients takes full advantage of this trend, enabling us to recapture nutrients before they vanish into the oceans.

GREEN FARMS

To produce the volume of food required to feed humanity by 2050 will demand massively more nutrients, whether on broadacre or smallholder farms. One way to obtain these is to stanch the losses occurring on the farm—to get far more food out of the fertilizer we already have. This is difficult but achievable.

To achieve this, first, it is important that pricing send the correct signals to farmers and that government subsidies not encourage wasteful and polluting fertilizer use. Second, it is important to educate farmers about the limitations of fertilizers—that no matter how much of a particular nutrient you pour on the crop, it will do no good if another essential nutrient is missing: the need for balanced, sparing fertilization is paramount. Third, there are many ways the farmer can better manage fertilizer use and distribution on the farm to avoid or limit losses—including using fertilizer planners that target the key growing times of the crop, using pinpoint application of the minimum amounts needed, choosing the right climatic conditions, analyzing soil so only the depleted areas are fertilized, using precision farming, using vegetation belts to intercept runoff and nutrients leaching through the soil, using topographic planning, recycling dairy and piggery waste, and so on. China, for example, has 1,830,000 pig farms that generate 1.2 million tonnes (1.3 million tons) of solid waste and 6 million tonnes (6.6 million tons) of liquid wastes daily.[24] Nine-tenths of this potentially valuable resource is currently lost—dumped in waterways and on land where it creates acute pollution and health hazards. Scientists are now working on ways to safely recycle this rich resource as fertilizer. Perhaps the largest and most pressing task is to disseminate these tools quickly to nearly two billion farmers around the world, a challenge we will explore later.

Finally, there is a need for an urgent worldwide scientific effort in soil biology to enable us to use microbes to unlock inaccessible nutrients in the soil. Much of the applied fertilizer and many naturally occurring nutrients in the soil are locked up, unavailable to plants. A deeper insight into the physical and microbial processes of various types of soil holds enormous potential for releasing these nutrients and raising the yields of crops and pastures worldwide—and some consider it a possible source of the next great leap in agricultural productivity. Soil biology as a field, however, has long suffered from underinvestment and a dearth of recruits. This is a situation urgently in need of amendment.

NOVEL FOOD SYSTEMS

One of the easiest ways to lower our dependency on artificial fertilizers would be to voluntarily reduce our intake of meat, dairy, and other livestock products. The University of Manitoba's Vaclav Smil has calculated that if affluent consumers lowered their intake of these foods by 15–35 percent, it would save between 5 and 15 percent of the phosphorus now used to grow the grain crops that feed the animals.[25]

In view of the looming shortages of land, water, and energy and the impact of climate change, however, it is important to recognize that traditional methods of farming used for centuries may not be able to produce food for all humanity sustainably into the future, and certainly not in all regions of the world. As a result we may have to develop alternative food-production systems that make far less use of land, water, and energy and rely instead on recycled urban nutrients, smart technology, and human labor.

Such systems already exist. Scientists have long grown cultures of plant and animal cells and microbes in the laboratory, and this procedure is capable of being scaled up to produce food in large vessels known as bioreactors. Just add warmth, water, and the right nutrients and you can produce edible, nutritious food by the ton from microbial, vegetable, or fungal cultures. Although this may sound fairly distasteful to the gourmet, one should never forget that fine wines, cheeses, salamis, and beers are all the products of microbial processes. In any case, many of today's savory snacks—especially sausage (into whose mysteries few dare to inquire) and surimi seafood—are made from recycled food "wastes," so this is merely an extension of the transformations the food manufacturing industry already performs. The chief argument in favor of this form of food production is its ability to convert hitherto wasted water and nutrients into a healthy, even interesting, diet with a minimum of energy, transportation, and other inputs. Such processes have long been contemplated for supporting colonies on the moon or Mars, but it may transpire that they will be needed in order to sustain spaceship Earth a little sooner.

A tantalizing new form of food production is "artificial photosynthesis"—artificially mimicking what plants do naturally, which is use sunlight to convert water and carbon dioxide into sugars and carbohydrates.[26] Teams of scientists around the world are already working on this challenge, which is primarily seen as a way of producing sustainable energy but can also, potentially, be used to produce sustainable food, using as a feedstock the carbon dioxide emissions from power stations or even vehicles. Although the idea of eating food made from your car's exhaust sounds unappetizing, it is no more so than eating the recycled nutrients from human waste, which we have been doing since the dawn of civilization. At least in theory, artificial photosynthesis offers the potential to produce nutritious food in large volumes from a limited area and by reusing waste carbon emissions.

A GLOBAL NUTRIENT PLAN

A global plan for tackling the looming nutrient crisis is no less essential and far more pressing even than a plan to deal with climate change. Yet it is not on the agenda.

To some extent, the planetary machinery for global warming is now in motion and—owing to the long residence time of carbon dioxide in the atmosphere and lags in the Earth system—cannot be put into reverse on a timescale shorter than centuries, although it may perhaps be slowed down. The nutrient crisis, by contrast, will be on us within a generation and is completely avoidable, though it will require strenuous measures to stave it off. At the moment the words that best describe the attitude of world and civic leaders to this issue are ignorance and apathy, so any change must be driven by the people.

The mindless depletion and waste of the planet's nutrients would make no sense to any of our ancestors. At the time of our greatest numbers and demand for food, it makes no sense for us.

It is time to end the waste.

WHAT CAN I DO ABOUT IT?

The following are ideas for reducing your individual waste of nutrients.

1. Eat less meat, dairy, sugar, and oils; eat more vegetables.
2. Favor foods produced using low inputs of artificial nutrients or best-practice use of fertilizers. Support food labeling of this.
3. Waste less food and compost all plant material.
4. Support large-scale recycling of urban organic waste into food production.
5. Use more compost and less fertilizer in your home garden.
6. Declare your independence: install a composting toilet.

SIX

TROUBLED WATERS

> Give a man a fish and he will eat for a day. Teach a man to fish and he will eat for a lifetime. Teach a man to create a shortage of fish and he will eat steak.
>
> **—Jay Leno**

During a brief frenzy in the late 1960s and early 1970s, Soviet and Japanese fishing vessels plundered more than a million tons of a strange, spiny fish called the North Pacific armorhead off the peaks of sea mounts, deep below the surface of the northern Pacific Ocean. When the fish ran out, the Soviet trawlers chugged off in search of fresh prey, discovering a delicious brightly colored goggle-eyed fish called the orange roughy, which also hung around sea mounts in deep waters south of New Zealand. Their find sparked a gold rush as trawl fishermen worldwide began to hunt down the roughy hot spots.

By 1990 the Australian roughy fishery was at its climax. Fifty-six boats lined up, often one behind the other, to target areas not much larger than a football field, hauling 62,000 tonnes (68,000 U.S. tons) of beautiful deep-red fish in a single year out of the lightless depths south of Tasmania, for a market that couldn't get enough of their succulent white flesh. Vessels were returning to port with their gunwales nearly awash in up to a million dollars' worth of fish crammed in their holds. "I know what a gold rush looks like," an older, wiser Stuart Richey, who was among the fishermen, reflects. "We were landing *boatloads* of fish. I know how it affected everyone's thinking at the time but, looking back, you can see how illogical it all was. It was a new, deepwater species just waiting to be caught—and we didn't have the correct science to guide us or the management to adequately restrain catches, so we just went out and caught them."[1]

Not realizing that the densely packed fish were spawning and not knowing that individual fish living in the unchanging, pitch-dark world of the deep ocean were 100–150 years old, the trawlers hammered in with all the firepower of modern fish-finders, satellite navigation, and sophisticated nets. Wealth poured into fishing communities that had been struggling to survive. In the two years that followed, the catch halved, and then halved again. By 1996, it was down to 5,000 tonnes (5,500 U.S. tons) and fisheries managers were panicking. A few years later the fishery was effectively shut, and a decade later it showed only a flicker of recovery. The boom had turned spectacularly to bust. Not just off Australia but off Chile and Namibia, in the northeastern Atlantic, and in the southwestern Indian Ocean similar collapses followed.[2]

"Why did fisheries science and management fail for the orange roughy?" asks Tony Koslow, a fisheries scientist who studied the tragedy as it unfolded. "A number of factors appear to be responsible, and they are worth exploring, if only because deepwater fisheries continue to collapse around the world. The first strategic error that doomed the Australian orange roughy fishery, among others, was violation of what I call the Scott Joplin principle. . . . Joplin often marked the tempo of his pieces, 'Not fast.' Later in life, he seems to have lost patience, writing above one of his scores: 'Notice: It is never right to play Ragtime fast. Author.' For *ragtime,* substitute *fisheries.*" Fisheries need to be played slow, Koslow concludes. The second error was to allow fishing to continue on a spawning population. And the third was a simple failure by fisheries managers to heed scientific advice, and their caving in to political pressure from the fishing industry.[3]

Koslow's point is that it is very easy with modern technology to exhaust a fishery if you don't know how big it is, its rate of replacement, or its food sources—and it is very hard to build it back up again. Time and again in recent history, this maritime "tragedy of the commons" has taken place. Decades after the destruction of the Grand Banks cod stocks, the most fecund fish in the oceans has still not recovered. Absolute caution is needed in developing any new fishery—but almost never is it exercised.

Worldwide, the evidence is mounting that the fish are running out. Almost one in three sea fisheries has collapsed or is in the process—just like the armorhead, the orange roughy, and the cod.[4] Most of the continental shelves have been swept clean and even miles down, in the deep ocean, the rapine is now taking its toll in a world where little has changed in millions of years. Incredible though it may appear, so insatiable is the human hunger for protein that we already appear to be mining the oceans to their limit—with incalculable consequences as world demand for food doubles and doctors and dieticians urge the affluent to devour more fish still for the sake of their overfed arteries and fatty hearts.

This ought not to surprise us. Each time the industrialized urban world embarks on a wild harvest, it almost invariably destroys the natural stock. In a powerful little book, the Greenland politician Finn Lynge once explained that when Inuit and other Arctic peoples hunted seals for thousands of years, the seals were in no danger of vanishing. But when city people demanded seal fur, seals' survival was threatened.[5] The lesson is similar for whaling: indigenous whaling rarely makes much impact on a stock, but industrial whaling can devour it utterly. And, with a third of the world's forests already gone, the rule seems to apply to unrestrained logging also. Behind the plunder is the insatiable, unthinking hunger of the city and its inhabitants for the finite products of a wild world with which they long ago lost contact.

This hunger today consumes around 140 million tonnes (154 million U.S. tons) of fish a year, 100 million tonnes (110 million tons) in wild catches and 40 million tonnes (44 million tons) in farmed fish. More than a quarter of this total, however, is wasted and thrown away in capture, processing, and preparation. In addition, two Australian academics have calculated, the world's pet cats consume around 2.5 million tonnes (2.75 million U.S. tons) of fish annually—meaning that in rich countries the average cat eats more fish in a year than the average human does.[6]

"If fishing around the world continues at its present pace, more and more species will vanish, marine ecosystems will unravel and there will be 'global collapse' of all species currently fished, possibly as soon as midcentury, fisheries experts and ecologists are predicting," the *New York Times* reported in November 2006. The *Times*'s report was based on the work of the Canadian scientist Boris Worm of Dalhousie University and thirteen other marine researchers, who reached their conclusion after extensive perusal of many scientific papers and Food and Agriculture Organization (FAO) statistics covering the period 1950–2003. According to the *Times,* the team found that by the early twenty-first century, 29 percent of the world's fish stocks had been pounded so heavily by fishing, pollution, or loss of habitat that they had collapsed—by which the scientists meant reduced to 10 percent or less of their original levels.[7]

There was a twist to the oft-told story of overfishing, however: the loss of fish was now so serious that it was actually undermining the health of the oceans, the researchers argued. "We conclude that marine biodiversity loss is increasingly impairing the ocean's capacity to provide food, maintain water quality, and recover from perturbations. Yet available data suggest that at this point, these trends are still reversible," they added. Worm pointed out that elephants and tigers are still around because humanity has taken steps to preserve them—and the same could apply to fish.[8]

The researchers' report unleashed a storm of criticism, both from scientists who challenged the method by which it was compiled and from fishermen and fisheries managers who disputed its conclusions. U.S. fisheries scientists branded it "inaccurate and overly pessimistic," arguing—somewhat obscurely—that falling catches didn't necessarily equate with falling fish numbers.[9] Worm was accused of headline-hunting, despite the fact that his papers appeared in some of the world's most prestigious scientific peer-reviewed journals. Historically, however, such reactions have erupted almost every time a fisheries scientist has spoken out about overexploitation of fish—and historically, too, warnings have usually proven to be conservative.

A World Bank study estimated the total economic loss from the world's fishing industries at $2 trillion over thirty years, including the run-down in "capital stock" of fish. It urged a global reduction in fishing effort, stronger marine "land rights," and fairer sharing of the benefits. Scientific studies have confirmed that where fishermen have well-defined rights to a fishery they are much less likely to overfish it.[10]

On the positive side, there is also now encouraging progress in a number of fisheries worldwide to put catches on a more sustainable footing, balancing the take with the rate at which the fish population can recover. "What is often overlooked is the progress that has been achieved by the thousands of dedicated fishery scientists and well-established national and international programs in place to assure sustainable management of the oceans," commented Thor Lassen, the president of the Ocean Trust, a fishing lobby. "The recovery of North Atlantic swordfish, New England groundfish and scallops; the success of cooperative industry management and research partnerships in the North Pacific and Atlantic fisheries; and the remarkable progress with endangered sea turtle recovery and reduction of shrimp fishery bycatch in the Gulf of Mexico demonstrate that our system of management is working." In Australia's scallop industry, characterized by almost a century of wild booms and busts, stability is at last within sight as fishermen and scientists combine forces to assess the stock and impose strict catch and area

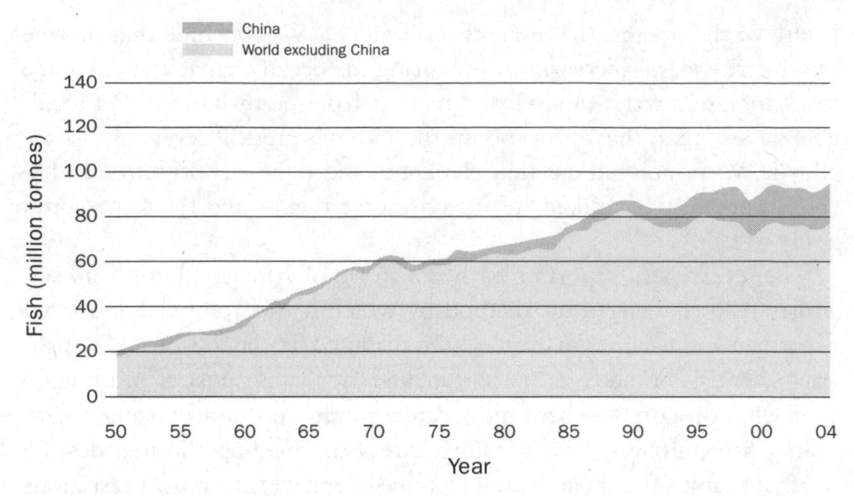

Figure 3 World wild fish catch, 1950–2004. All countries except China show a decline. Source: Food and Agriculture Organization of the United Nations, "World Fish Production, China and the Rest of the World," fig. 29, in *The State of Food and Agriculture 2006* (Rome: Food and Agriculture Organization, 2006), ftp://ftp.fao.org/docrep/fao/009/a0800e/a0800e.pdf.

limits based on the shellfish's ability to recover from harvesting.[11] Similar developments are occurring worldwide as fishermen move from being hunters and gatherers of the seas to husbanding and managing their resources.

Global fishing figures from the FAO showed that the world's total wild fish catch had been stagnant for fifteen years (1989–2004), while aquaculture had undergone steady expansion to the point where it was furnishing two-fifths of the total fish supply in 2004. The data did hint, however, that the world catch was falling off, offset only by frenetic and probably unsustainable Chinese fishing activity (figure 3).[12]

All this indicated that the world's four million fishing boats and forty-one million fishers were competing harder, smarter, and longer for fewer fish. If the total haul hadn't actually peaked in 2000, as the figures suggested, then "peak fish" certainly appeared imminent. The FAO's appraisal was that a quarter of the world's fish stocks were underexploited, half were fully exploited, and the rest were overexploited, and this situation appeared to have persisted for a decade or more. "The maximum wild capture fishery potential from the world's oceans has probably been reached," it stated. Within this, however, stocks of the big meat-eating fish (tuna, billfish, snappers, dories, and the like) were dwindling and fishermen had been forced instead to target the smaller grazing fishes such as anchovies, pilchards, and sardines.[13]

Worm's research had revealed that the most critical region lies south and east of Asia. Indeed, the former director general of WorldFish, Meryl Williams, has warned that Southeast Asia's seas are running out of fish, putting the livelihoods of up to one hundred million people at risk and increasing the likelihood of piratical incursions and hostile disputes. "These fisheries have expanded dramatically in recent decades, and Indonesia, Thailand, Vietnam and the Philippines are now in the top twelve fish producing countries in the world. In the Gulf of Thailand, Thailand's most important fisheries location, the density of fish has declined by 86% from 1961 to 1991," she says. "Between 1966 and 1994, the catch per hour in the Gulf by trawlers declined more than sevenfold. In Vietnam . . . the total catch only doubled despite a tripling in the capacity of the fishing fleet. In the Gulf of Tonkin, where resources are shared with China, the fish catch per hour in 1997 was only a quarter of that in 1985. In the Philippines, most marine fisheries were overexploited by the 1980s, with catch rates now as low as 10% of rates when these areas were lightly fished."[14]

The Asian fisheries experience is important, because that is the region of the world in which food demand, especially for protein, has soared dramatically in recent decades—and where the most fights over fish occur. It presages what is likely to happen worldwide as humanity's food requirements double approaching 2050. "By 2020 developing countries will account for nearly 80 percent of fish production and consumption," predicted Joachim von Braun and colleagues, of the International Food Policy Research Institute. "Projections . . . show that fish consumption in developing countries will increase by 57 percent between 1997 and 2020 as a result of rapid population growth, increasing affluence, and urbanization. Fish farming, or aquaculture, already a booming industry, will continue to expand in order to meet more than 40 percent of this demand, since most wild fisheries are tapped to capacity or beyond."[15]

One thing the fish catch numbers clearly point to is that there probably aren't another 100 million tonnes (110 million U.S. tons) of fish in the oceans easily catchable to meet the doubling in world food demand. Despite the strongest growth in fishing effort in all of history, fish catches remained stubbornly stagnant, perhaps even falling slightly. If this remains the case, the extra 100 million tonnes of fish meat protein not obtainable from the oceans may have to come from one of two other sources: aquaculture or land-based livestock.

Suppose for a moment that for all of this fish we substituted meat from the land-based farming of poultry, pigs, sheep, goats, or cattle. This would require an extra 1,000 cubic kilometers (250 cubic miles) of freshwater, either rainfall or irrigation, to grow the crops to feed the animals—and this would be on top of the 1,850 cubic kilometers (450 cubic miles) required to meet the growth in demand for meat anticipated by the FAO. So, by 2050, increases in meat production alone could consume a volume larger than all the irrigation water used worldwide in farming today—at a time when cities will be seizing half of the farmers' water, rivers will be dying, and the climate will be drying. Another way to read the equation is in the form of grain: at 10 kilograms (22 pounds) of grain, roughly, for each kilogram (2.2 pounds) of meat produced, the world will need to grow an extra billion tonnes of grain simply to feed the animals needed to substitute for the fish we will no longer be able to take from the seas or rivers, *even if today's catches were to be sustained into the future.* That would require almost the equivalent of a doubling in the North American coarse grains harvest. Where all that extra grain, and the farmland, water, and fertilizer needed to grow it, will come from is not clear.

FISH FARMING

The farming of fish and other aquatic life, including algae, offers promise for expansion. Its output has risen strongly in recent decades to the point where by 2008 it was supplying about 40 million tonnes (44 million U.S. tons) of the world's 140 million tonnes (154 million U.S. tons) of fish and other aquatic products. Altogether, about 240 different freshwater and saltwater species are now being farmed around the world, and industry growth has averaged about 10 percent a year since 1970.[16]

The things that constrain aquaculture, however, are similar to those that limit land-based farming—lack of suitable land or coastline for ponds, polluted water, environmental regulation, land-tenure issues, sea-level rise, and the availability of feed supplies. It takes roughly five tons of wild-caught fish to make fishmeal to feed a single ton of farmed fish, so fish farming can be even more destructive of marine fish stocks than actual fishing. Consequently, more and more farmed fish, like chickens, pigs, or feedlot cattle, are fed on grain. This competes with other uses for the grain, including human food, and imposes greater stress on the agricultural landscapes called on to grow it. In other words, farming fish also degrades the land. At the same time, the quality of water needed to grow fish or prawns in is declining disastrously in many countries, as these countries dump toxic industrial wastes, pesticides, oil, hormones, heavy metals, nutrients, and sewage into their waterways. Fish obligingly concentrate these poisons in their meat and pass them on to the humans who eat them. Aquaculture itself produces pollution—in the form of nutrients and the antibiotics and fungicides needed to keep farmed fish healthy; the growing reluctance of local authorities to permit fish farms in areas with sensitive waterways is yet another brake on the industry's future growth.

A scientifically controversial option is to seek to replenish wild fisheries by breeding and rearing hatchlings artificially and then releasing them to sea. This has long been done with salmon, and recent breakthroughs in raising tuna, abalone, and other species from the egg make this form of extensive "ocean ranching" more than just a possibility as a way to sustain dwindling wild populations; however, the effect of this reseeding on natural populations, the wider marine environment, and other species is unknown.

An important development is that the world harvest of aquatic plants has, for the first time, reached 15 million tonnes (16.5 million U.S. tons). This is far short of the 2.5 billion tonnes (2.8 billion U.S. tons) of grain we grow on land, but it is a vital signpost to the farming of sea plants as a major future source of food and feed, which has been neglected in most countries for too long. Sea plants, like fish, depend on good water quality, however.

On the intensive farming front, fish are extremely efficient at turning grain into meat—far more so than livestock or even poultry, which implies that if grain supplies are scarce, fish may well be more efficient to produce than pigs or cattle, provided clean water is available. Integrated systems such as aquaponics—the raising of fish and vegetables together in water tanks, with the fish fertilizing the plants and the plants cleaning the water—appear promising. Nonetheless, although she is positive about the potential for aquaculture to expand in the future, Meryl Williams questions whether the global industry will be able to grow fast enough to supply the unfulfilled demand for marine protein, should wild harvests continue to stagnate or decline.[17]

ACIDIC OCEANS

The most profound impact of human activity on Earth may be what we're doing to the oceans: turning them slowly but surely more acidic. Each time we start a car, use coal- or gas-fired electricity, or travel by plane, train, or boat, half the carbon dioxide emitted from the burning of the fossil fuel ends up in the sea—25 billion tonnes (27.5 billion U.S. tons) of it per year. Each molecule of carbon dioxide turns the ocean imperceptibly more acidic, in a process that is happening regardless of global warming or other changes.

Researchers have measured observable changes in the ocean's acidity. They have also demonstrated that even small shifts can kill corals, sponges, and certain common marine algae and plankton that are a foundation of the entire oceanic food web, including all its fish.

"The oceans are absorbing carbon dioxide (CO_2) from the atmosphere and this is causing chemical changes by making them more acidic (that is, decreasing the pH of the oceans)," explains Britain's preeminent scientific body, the Royal Society. "In the past 200 years the oceans have absorbed approximately half of the CO_2 produced by fossil fuel burning and cement production. . . . Ocean acidification is essentially irreversible during our lifetimes. It will take tens of thousands of years for ocean chemistry to return to a condition similar to that occurring at preindustrial times (about 200 years ago)." The Royal Society went on to warn, "Our ability to reduce ocean acidification through artificial methods such as the addition of chemicals is unproven. . . . Reducing CO_2 emissions to the atmosphere appears to be the only practical way to minimise the risk of large-scale and long-term changes to the oceans."[18]

According to Malcolm McCulloch, an earth scientist with Australia's ARC Centre of Excellence in Coral Reef Science, recent research into corals has found that the ocean has become more acidic since the 1960s. "This is still early days for the research, the trend is not uniform and we can't as yet say how much is attributable to human activity—but it certainly looks as if marine acidity is building up," McCulloch said. "It appears this acidification is now taking place over decades, rather than centuries as we originally thought. It is happening even faster in the cooler waters of the Southern Ocean than in the tropics. It is starting to look like a very serious issue."[19]

Corals, algae, and plankton with chalky skeletons are at the base of the marine food web. They rely on seawater saturated with carbonates and bicarbonates to form their structures, just as we use calcium to form our bones. As more carbon dioxide dissolves out of the air and acidity rises, however, the carbonate saturation of seawater declines, making it much harder for these animals—and indeed all shellfish—to calcify, or grow their shells and skeletons.

"Analysis of coral cores shows there has been a steady drop in calcification over the last 20 years," says Ove Hoegh-Guldberg, a coral authority at the University of Queensland. "There's not much debate about how it happens: put more CO_2 into the air above and most of it dissolves into the oceans. When CO_2 levels in the atmosphere reach about 500 parts per million (ppm), you put calcification out of business in the oceans," he warns. The world's atmospheric greenhouse gas levels are presently about 450 ppm carbon dioxide equivalent. Even with major efforts to cut greenhouse emissions, they are expected to reach 550 ppm by midcentury, driven by industrial growth in China and India and continued expansion in fossil fuel use everywhere. At such a level, Hoegh-Guldberg fears, the world's coral reefs will simply die.[20]

"It isn't just the coral reefs which are affected—a large part of the plankton in the Southern Ocean are also affected. These drive ocean productivity and are the base of the food web which supports krill, whales, tuna and our fisheries. They also play a vital role in removing carbon dioxide from the atmosphere."[21]

That a global loss of corals is possible is hinted at by the most catastrophic event ever to occur in the history of life on Earth—worse even than the extinction of the dinosaurs. Known from the fossil record as "the Great Dying," this extinction occurred at the end of the Permian era, 251 million years ago, when, geologists say, it eliminated 96 percent of all sea life, including most of the fishes. Scientists still argue over the precise causes of the Permian extinction, but a likely suspect is a vast outbreak of volcanic activity in Siberia at about that time, which belched so much carbon dioxide and sulfur into the atmosphere that the seas turned sharply acidic. As millions of dead organisms rotted, the oceans were filled with vast bacterial blooms that stripped their oxygen, the way highly polluted waters turn anoxic today. As the oceans warmed, they also stratified, producing huge anoxic dead zones. These effects killed off the fish, trilobites, nautiloids, and other organisms that had not already succumbed to acidity. It is this combination of acidic water and lack of oxygen that scientists fear could be lethal to corals, plankton, and fish as manmade pollution increases.

The coral authority Charlie Veron notes that there have been five mass extinctions that either wiped out or partly eliminated the corals and most of the Earth's sea life at the time. In each case, high carbon dioxide levels and acidic oceans are thought to have played a key role. And in every case it took ten million years or longer for the ocean equilibrium to recover and for corals to appear again in the fossil record.[22]

"The prospect of ocean acidification is frightening," Veron argues. "It is serious because of *commitment*—a word that will soon be used with increasing frequency in the scientific literature." Commitment, essentially, means that a process is unstoppable. If the oceans turn acidic—as they are already doing—the only known way to reverse this is the slow weathering of limestone mountain ranges into the sea, a process that takes millions of years. "It cannot rationally be doubted that we are now at the start of an event that has the potential to become the Earth's sixth mass extinction. This time there are no bolides (asteroids), no supervolcanoes, and no significant sea level changes. . . . It is a case of humans changing the environment," Veron adds.[23]

The lethal combination of acidic oceans, coral bleaching caused by warming of the oceans, and human destruction is already taking its toll on the world's reefs. "Corals are dying out around the world. . . . In the Caribbean, live coral cover has fallen from an average of [about] 55% in 1977 to 5% in 2001," writes Jeremy Jackson of the Scripps Institution of Oceanography in La Jolla, California.

> Synergistic effects of habitat destruction, overfishing, introduced species, warming, acidification, toxins, and massive runoff of nutrients are transforming once complex ecosystems like coral reefs and kelp forests into monotonous level bottoms, transforming clear and productive coastal seas into anoxic dead zones, and transforming complex food webs topped by big animals into simplified, microbially dominated ecosystems with boom and bust cycles of toxic dinoflagellate blooms, jellyfish, and disease. Rates of change are increasingly fast and nonlinear with sudden phase shifts to novel alternative community states. Halting and ultimately reversing these trends will require rapid and fundamental changes in fisheries, agricultural practice, and the emissions of greenhouse gases on a global scale.[24]

Such catastrophic trends will first affect the five hundred million people who depend directly or indirectly for their livelihoods on coral reefs because, when reefs die, the fish disappear too. These people will be forced to obtain their food from the land. Then it will affect the rest of us, in refugee crises, rising prices, and increased scarcity of fish. The coastal barriers formed by reefs will crumble, exposing more seaside towns to hurricanes, tidal surges, and tsunamis. The oceanic food-chain impacts are quite unclear as yet, but will probably mean reduced numbers of certain classes of phytoplankton—the most numerous form of life on Earth—and their replacement by others, with corresponding effects all up the food chain to fish.

We are changing the oceans as profoundly as we are changing the air we breathe, risking an important component of our future food supply—and the only known way to prevent this is for humanity to "decarbonize," to stop emitting carbon dioxide.

SOLUTIONS

The first and most practical thing that can be done to arrest the decline of the world's wild fisheries is to restrict catches until the dynamics of fish populations are better understood and they can recover from what is taken. This can be achieved through modern fisheries management methods that involve assessing both the fish stock and its ecosystems and limiting the fishing effort, with the full support and active cooperation of the fishermen, to what is judged sustainable. The process is well understood—but rarely implemented. In some developing societies it can be achieved by respecting and reinforcing the traditional rules of the sea decreed by elders, which are often based on a wish to husband the village's precious food resources. Coupled with this, a global war should be declared by all governments on pirate fishing, which is wantonly mining fish stocks and destroying people's livelihoods in both territorial and international waters, and rapacious methods such as dynamite- and cyanide-fishing and the live-fish trade should be banned. Such measures may be able to stabilize some fisheries that are not already overexploited. At the same time there must be rigorous assessment of the stock and ecology of any proposed or underdeveloped fisheries, to avoid future overexploitation errors.

A factor demanding urgent attention is waste: almost a quarter of all the fish caught—25 million tonnes (27.5 million tons) per year—are thrown away. Often these are fish of low market value known as bycatch or, contemptuously, as "trash fish." Yet they all represent protein and potential food for people or feed for animals, and it is a great shame that they should be killed for nothing. So, in conjunction with a global movement to prevent the waste of food between farm and the consumer's fork, we need measures to prevent or recycle this prodigal wastage of the ocean's bounty. In advanced fisheries, special nets that exclude nontarget species (or allow them to escape) and ways to add value to low-value fish by turning them into delectable seafood morsels are already being tried, but these need to spread worldwide as a matter of urgency.

Fishing is an efficient industry in the sense that the fish do all the hard work of gathering nutrients for us—and we only have to gather them. So more scientific effort must be put into understanding how oceanic and local fish stocks can be renewed, perhaps by assisted breeding or redistribution, as well as by natural replenishment and the management of marine food chains. Also, because intensive aquaculture is always risky in terms of pollution or disease, greater effort is needed to pioneer extensive forms of mariculture—a kind of "pastoralism of the seas" in which large marine ecosystems, entire bays, bights, and currents are managed with a view to maximizing production of food in the same way that large cattle or sheep enterprises are managed in the world's rangeland areas.

Just as humans have progressed from hunting animals to developing a food system based on farming crops over the past ten thousand years, we need to apply similar thinking to the oceans. The growing of sea and water crops (calling them "seaweed" perpetuates the foolish idea that they are of no value) will probably underpin an effective and sustainable aquatic harvest in the future. A huge worldwide effort is needed to devise means to efficiently grow and harvest water plants, both as food for us and feed for fish and land animals. Indeed, water plants are one way we can reharvest the nutrients that we unthinkingly pour into bodies of water. If this is achieved, there is little doubt we can greatly expand the amount of food we obtain from the seas, rivers, and lakes. If we continue to rely only on aquatic "livestock" like the hunters of old, it is probable that our burgeoning demand will far outrun wild fishes' ability to replenish themselves. We will exterminate big fish as surely as we did the megafauna of Europe, Asia, North America, and Australasia, and almost did the great whales. The world's unmet demand for protein will then have to come entirely from overstressed land-based farming systems.

The ultimate power to improve the world's fishing industries and to run pirates and plunderers off rests with the urban consumer—and so does the responsibility. Governments may regulate: consumers decide. Thus, what is needed most is worldwide education to help consumers to discriminate between fish that are produced by sustainable means and those that are obtained by mindless, greedy plundering. Have the fish in tanks in your local seafood restaurant, for example, come from well-maintained farms with a light environmental footprint, or were they caught by poor fishermen using cyanide sprayers under the direction of a ruthless and uncaring pirate company? Are pilchards more sustainable as a food choice than dories or roughies? How do you feel about eating a deep-sea fish that may be four times your age—and is it sustainable to do so? If you would reject a seal fur coat or whale meat, why would you not also reject a similarly unsustainable fish product?

The debate over the state of health of the world's fisheries and oceans will continue, and to some extent this acts as an impediment to our tackling the issue. If we do not accept that there is a problem, we will have difficulty in mustering either the will or the resources to solve it. The role played by the oceans in our efforts to double the world's food supply by midcentury could thus be far below their full potential—indeed, they could exacerbate the coming famine, instead of helping to avert it.

WHAT CAN I DO ABOUT IT?

1. Eat fish sparingly and ensure that none is wasted.
2. Be an informed and selective consumer, buying fish from wild fisheries that can demonstrate that they are sustainable—that is, where the catch is balanced by natural replacement.
3. Choose grazing fish in preference to the larger predatory fish, which are needed to maintain ecosystem balance.
4. Choose farmed fish, but demand information to show that its production processes are clean, safe, and sustainable.
5. If you need more omega-3 oils in your diet, then consider algal sources or even crops enhanced to produce these. These oils originate with plant-life in any case, not with fish.
6. Support government policies that conserve water, protect its quality, and keep it free from toxic contamination so it can be used to grow fish.
7. Support all measures that will eliminate greenhouse gas emissions, to reduce the risk of ocean acidification. Reduce your own emissions.
8. If you are a fisher, practice catch-and-release.

SEVEN

LOSING OUR BRAINS

> Almost no country has achieved a rapid ascent from hunger and poverty without raising agricultural productivity.
>
> **—Bill and Melinda Gates**

The killer exploded out of eastern Africa a year before the millennium. Stealthily, traveling on the wind and attached to people's clothing, it spread from one country to the next in a dominolike succession: Uganda, Kenya, and then Ethiopia and Sudan. Clearing the Red Sea in a single bound, it entered Yemen. By 2008 it had hurdled the Arabian Gulf, had invaded Iran, and was poised, like an angel of death, on the borders of Pakistan and India. Today many people have still not heard of Ug99. Yet it has the potential to kill more people than bird flu or AIDS, inflicting vast disruption on the world economy—and a pandemic may already be unstoppable.

Ug99 is a rust, a brownish fungus that attacks cereal plants. But it is a strain of rust more deadly to humankind's staple food, wheat, than anything yet seen. Potentially wiping out entire crops, researchers have established that as many as 85 percent of the world's wheat varieties may be susceptible to it. People have forgotten how devastating such events can be: the last major outbreak of stem rust occurred in North America in the early 1950s, when another, less deadly, strain devoured two-fifths of the continent's spring wheat crop.[1] But that was long before the world population grew so large, transportation so pervasive, or economies so interwoven. Ug99 is—at least potentially—the Irish potato blight of the

1850s on a global scale, with the capacity to bring famine and death to tens of millions.

No one knows where Ug99 came from. It was first discovered in a plant nursery in Uganda in 1999, hence the name, but scientists suspect that it originated somewhere in southern Africa when a conventional rust underwent a tiny mutation that suddenly enabled it to attack three-quarters of the world's wheat crops. It tore loose in Kenya, where it destroyed up to 80 percent of the wheat harvest because most of that country's wheat plants had only a single gene protecting them against rust. Indeed, the world's wheat crops are mainly protected by just three antirust genes—and Ug99 has broken through against all of them. "This is a global threat," Masa Iwanaga, then the director general of the International Wheat and Maize Center (CIMMYT) told a world scientific crisis meeting. "The risk of a stem rust epidemic in wheat in Africa, Asia and the Americas is real, and must be averted before untold damage and human suffering is caused," said Mahmoud Solh, the director general of the International Center for Agricultural Research in the Dry Areas.[2]

Scientists were pulled off their vital work of trying to feed the world in the coming generation and refocused on trying to dam this sudden and disastrous breach that had opened in the security of the global food supply. An international scientific task force was formed, headquartered at Cornell University and funded by the Bill and Melinda Gates Foundation, among others, as researchers rummaged through the genetic profiles of grain crops ancient and modern to see if they could pinpoint genes able to protect against the new blight and then breed them, three or more at a time, into the world's wheat varieties. It is a colossal undertaking and, even if the genes are quickly found and prove to be protective against Ug99, it will be years before even half the world's wheat is safe. Meantime, humanity teeters on the brink of a bread catastrophe.

In a sense, Ug99 was a disaster waiting to happen. As flu regularly reminds us, diseases mutate all the time—but become major killers only when the ground is fertile and society unprepared. For almost a generation, the attitudes of governments worldwide have ensured that the world is becoming less well armed than it could be against this and other aspects of the coming famine. By assuming that the food problem was solved, by eroding support for the global effort to feed ourselves, by trusting to the private sector to do it all for us, governments have prepared the ground for the food shortages of the mid-twenty-first century. Some may feel that this criticism of governments for their complacency and lack of foresight is unfair—but leaders who do not ensure that the people are fed fail the first test of leadership.

THE WAR ON HUNGER

Ten thousand years ago the first farmers gathered suitable grasses, transported them to the home site, and began to clear land to grow them. Thus they both selected the best genes for human food production and created the environment in which they could thrive. This simple innovation gave rise to civilization as we know it and from it farming rapidly expanded from less than a tenth of the Earth's land area in 1700 to more than 40 percent today. However bountiful the new source of food, it was also rather unreliable, affected by weather, disease, and competition from weeds, insects, and pests. The advent of scientific crop breeding in the late nineteenth century, based on the genetic discoveries of the Austrian monk Gregor Mendel, provided the first true counter to this unpredictability and began to deliver the more secure food supplies on which the rise of industrial civilization depended. Coupled with improved fertilization and methods of weed and pest control, yields of wheat—for example—rose from 1–2 tonnes (1.1–2.2 U.S. tons) per hectare in the late nineteenth century to around 7–8 tonnes (7.7–8.8 U.S. tons) in advanced farming countries with reliable climates by the early twenty-first. From the 1970s on, the international agricultural research centers of the Consultative Group on International Agricultural Research (CGIAR) spread similarly dramatic gains in yields of rice, maize, grain legumes, potatoes, and other crops throughout the developing world, enabling once-hungry nations such as India, China, and Mexico to grow all their own food.

Few, if any, of the quiet heroes who achieved these remarkable feats for us all are household names.[3] Perhaps the best known is the late Norman Borlaug, one of the fathers of the high-yielding dwarf wheats that are said to have nourished more than two billion people since their development, who received the Nobel Peace Prize in 1970 for his achievements. In 1962 M. S. Swaminathan—another giant among those who have fed the world—arranged to bring Borlaug and his wheats to India and began, quietly, to refute skeptics such as Paul Ehrlich, who in his 1968 bestseller *The Population Bomb* had argued that the country could not possibly feed itself: "The battle to feed all of humanity is over. . . . In the 1970s and 1980s hundreds of millions of people will starve to death in spite of any crash programs embarked upon now. At this late date nothing can prevent a substantial increase in the world death rate." Yet, by 1974, barely six years after the distribution of high-yielding wheat and rice varieties, India was meeting all its own basic food needs.[4]

When the Green Revolution began, one in three of the world's people faced hunger or died from the diseases associated with it. By the early twenty-first century this had fallen to one in eight (or 850 million)—still an unacceptably high toll in human suffering and a number that grew again with the food crisis and global recession. The Green Revolution's major achievements included

- spectacular increases in yields and production of rice in Asia, Latin America, and Africa,
- dramatic improvements in both quality and yields of corn,
- a major increase in the world wheat harvest,
- farmed fish that grow 60 percent faster,
- many new varieties of beans, lentils, cowpeas, and pigeonpeas that yield more and resist disease better,
- increased food security and reduced malnutrition for hundreds of millions in developing countries,
- a platform for economic development in countries such as India and China, and
- a return of about 17:1 on every dollar invested in research.[5]

A major food crisis had been averted and a road out of poverty created for billions. The Green Revolution was not without downsides, however, coming under fire from critics for its unintended impacts on the environment and on social inequity—issues that have been addressed in recent programs, which have a far greater focus on microeconomic factors and the needs of smallholders and women, who make up the majority of the world's farmers.

The outstanding success of the Green Revolution in dispelling the Malthusian nightmare of the late twentieth century in most regions other than Africa also contained the seeds of its own undoing—the paralyzing complacency and neglect of agricultural science and technology on which to build the next great leap in food output. Over the ensuing years, the public effort put into maintaining and increasing food production began to decelerate and, in some countries, even to contract.

For the past quarter century, the brainpower required to feed humanity has been shrinking in relation to the global population and its needs. In local field research stations, in national agriculture departments, in universities, colleges, and research centers, and in the international agricultural research endeavor, funding has been cut or allowed to erode, labs and field stations have been closed, and promising research programs have been terminated. Many of the scientists who fed the world have quit in anger, sorrow, or disappointment, have been fired, or have retired, while recruitment has fallen off. The powerhouses of agricultural knowledge—the United States, Germany, France, Japan, Canada, and Australia—have turned away from agri-science in pursuit of other technological El Dorados. A report by Alex Evans for Britain's Royal Institute for International Affairs says that between 1980 and 2006 the proportion of the world's aid budget spent on agriculture dwindled from 17 to just 3 percent. "Total aid spending on agriculture fell 58% in real terms over the same period," it added.[6]

Support for the global effort to lift food production in developing countries has been stagnant for thirty years (see figure 4). Even in the food-insecure giants China and India, research efforts in other fields have eaten into that devoted to agriculture. The resulting dilapidation in the enterprise that feeds the Earth has disheartened a generation of young would-be agricultural scientists, especially in developed countries where many universities and colleges of agriculture could not find enough students willing to fill the places they were offering.[7] A gaping deficit in the river of knowledge and technology on which farmers depend to maintain growth in food production has opened up, which could take decades to fill. Speaking of the causes of the food price crisis, Tom Lumpkin, the director general of CIMMYT, said: "We have ourselves to blame. Our leaders did not focus on the food security of their grandchildren. They only saw the next election campaign coming. The world has a real shortage of leaders who read history books and understand global trends. Now we don't have the research or training to do what is needed."[8]

Bearing out his words, funding for the CGIAR's sixteen international agricultural research centers that built the Green Revolution technologies has in real terms been flatlining for more than thirty years. Despite the apparent upward trend in investment, the *real* spending on international science to overcome world hunger and secure global food supplies was less than $100 million a year from 1976 to 2007—a scandalously inadequate sum, which in inflation-adjusted terms is scarcely more than

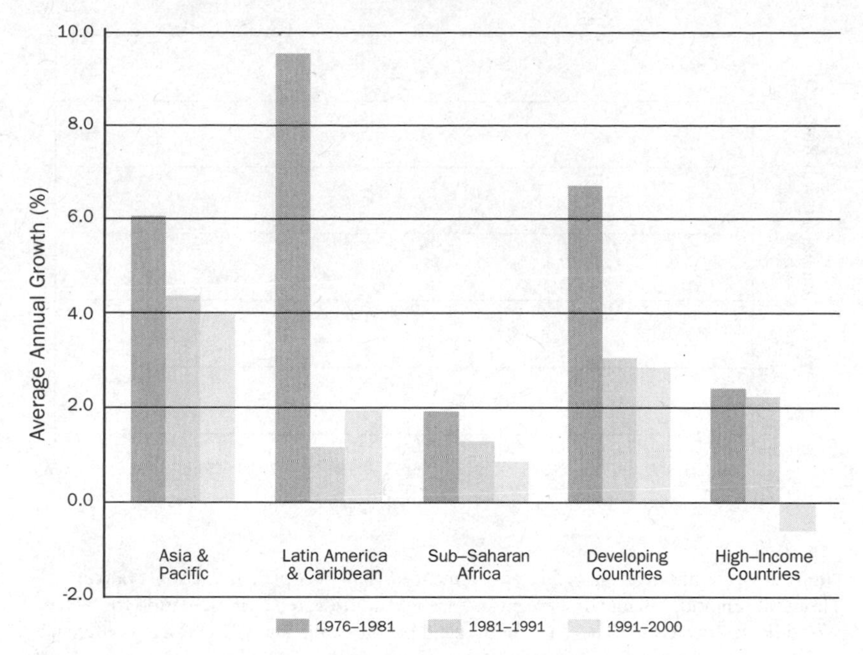

Figure 4 Declining trends in investment in agricultural science by region. Source: Philip G. Pardey et al., "Science, Technology and Skills," October 2007, table 6, p. 47, report commissioned by the CGIAR Science Council as a background paper for the 2008 World Development Report of the World Bank, http://siteresources.worldbank.org/INTWDR2008/Resources/2795087-1191427986785/PardeyPEtAl_SciTech&Skills–ALL1.pdf. Note: The category "Developing Countries" includes the data provided separately for "Asia and Pacific," "Latin America and Caribbean," and "Sub-Saharan Africa."

the sum invested was when the world had half the number of people and a far lower standard of living. "The lack of significant increases in real funding for the [international agricultural research] System, while it continued to grow and add more [research] Centers and more researchable areas, was accompanied by a simultaneous trend of declining funding for agricultural development in general over the same time period," a CGIAR funding task force reported.[9]

By the start of the twenty-first century, rich countries were spending barely 1.8 cents in every science dollar on agriculture, so unimportant had food become to them. China, India, and Brazil, which had invested 12 cents of their science dollar on food in the 1980s, reduced it to 7 cents as other research priorities took over. At the turn of the millennium, public investment by all governments worldwide in improving food production totaled just $23 billion[10]—a figure not large by the

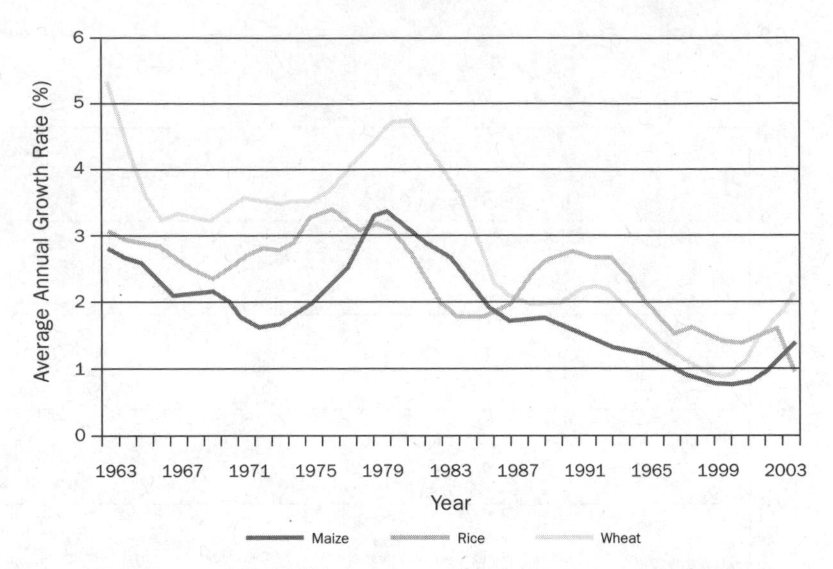

Figure 5 Declining trends in yield improvement of major crops, 1963–2003. Source: Thomas Lumpkin, "Blame Investment in Agronomic Research?" presentation for "The World Food Price Crisis: The Challenge Ahead for Australia and CIMMYT," conference in Canberra, September 2008.

standards of a single country, let alone an entire planet, and one that contrasts eerily with humanity's total spending of $1.5 trillion on armaments—as if killing one another were fifty times more important to us than eating. The neglect was most stridently displayed in the steady collapse of gains in grain yields (figure 5).

The agricultural economists Phil Pardey, Julian Alston, and their colleagues warn that the situation with regard to the funding of agricultural research is bad and getting worse: "Underfunding of agricultural R&D is pervasive, especially in developing counties. This trend is alarming given:

- the continuing and substantive growth of populations, especially in developing countries
- an increasingly scarce and deteriorating natural resource base
- the pervasive pockets of hunger and poverty that persist in developing countries, in many cases despite impressive national average productivity increases
- the growing divergence between developed country research agendas and the priorities of developing countries."[11]

Following the trebling in rice prices in 2007–8, the International Rice Research Institute (IRRI) commented: "Failure to act now through a wholesale reinvestment in agriculture—including research into improved technologies, infrastructure development, and training and education of agricultural scientists and trainers—could lead to a long-term crisis that makes the price spikes of 2008 seem a mere blip." Adding that progress in raising rice yields had fallen by two-thirds in recent years, the IRRI chair, Elizabeth Woods, stated, "The global community needs to remember two key things. First, that growth in agricultural productivity is the only way to ensure that people have access to enough affordable food. Second, that achieving this is a long-term effort. A year or two of extra funding for agricultural research is not enough. To ensure that improved technologies flow from the research and development pipeline, a sustained re-investment in agriculture is crucial."[12]

Standing against the tide of neglect, and departing from her prepared remarks, the World Bank's vice president for sustainable development, Katherine Sierra, told the Crawford Fund Conference on Agriculture and Climate Change in Australia that it was her aim to double global contributions to the international agricultural research centers, whose funding has been stagnant for the better part of three decades.[13] This, however, was before the collapse in the world financial markets and the onset of recession in major donor countries. In November 2009 the World Summit on Food Security meeting in Rome vowed to "reverse the decline in domestic and international funding for agriculture, food security and rural development," to "promote research for food and agriculture," to "reinvigorate national research systems," and to "improve access to knowledge," but in the absence of specific monetary commitments, clearly specified goals, and timelines, it was just words.[14]

To defeat the coming famine will take more than good intentions—and, notwithstanding global recession, the work must begin immediately. Indeed, since agriculture underpins so many countries' economies, it represents an important part of any sound strategy for economic recovery as well as for world peace.

PRIVATE VERSUS PUBLIC?

An important dimension in this global "slowdown" in agricultural research has been a shift from public- to private-sector investment. Policy makers often seem to assume it does not matter very much whether a piece of agricultural technology is invented in the public or the private domain—but the actual choice of technology is critical. Public research bodies tend to select technologies to work on that deliver public good outcomes, like putting more food on the world's table, helping the poor, improving health, or curbing the environmental impacts of agriculture, whereas private research tends to favor technologies that offer the greatest potential for profit to the investing company. There is also a presumption that it is farmers who drive the demand for these more profitable new technologies, whereas it is often agribusiness—meaning the corporate technology producers upstream of the farm and the food processors downstream of it—that decides which technology it wants and then exhorts or compels farmers to adopt it. This is not a criticism of agribusiness or of private-sector R&D but simply an observation that, just as a giant mining, retail, or computer company does not necessarily consider its primary job to be prevention of starvation or eradication of poverty, neither does a large agribusiness firm—which is equally answerable to its shareholders. Achieving global food security is seldom the private sector's first concern, and this needs to be borne in mind when decisions are made about the scale of public investment in food science.

To offset the trend somewhat, however, has been a steady rise in philanthropic funding, originating from the private and not-for-profit sectors but dispensed for the global public good. By the early twenty-first century, the Rockefeller Foundation, which, along with similar organizations, has strongly supported international agricultural research since the very start of the Green Revolution, had been joined by newcomers such as the Bill and Melinda Gates Foundation.[15] By 2009, the Gates Foundation was putting its money into everything from collecting seeds of African vegetables for selection and improvement to fighting the war on crop diseases such as Ug99, developing drought-tolerant maize, designing microirrigation systems for smallholders, gathering data on human hunger, and trying to understand the impact of development on small farmers. Its total contribution to the international agricultural research centers of the CGIAR in 2007 was $24 million, making it the organization's sixth largest donor—larger than most of the world's national governments.[16] The Gateses, clearly, could see the urgency of the food issue, to which the majority of governments and politicians were blind.

A further dimension of the public-private debate over agriculture has been the expanded use in advanced countries of intellectual property (IP) rights for new crop varieties and farming technologies, which, while invaluable in providing a financial incentive to make new discoveries and innovations, also has the effect of delaying or withholding those innovations from poor farmers (in all countries) who cannot afford to pay the royalties. As with the debate over the high cost of anti-AIDS drugs, it is questionable whether the developed world has a moral right to withhold farming technologies that may be the means of saving millions of lives—and, in view of the threat to global stability posed by food shortages, it is equally questionable whether it is common sense to do so. Imagine, for example, how the 1970s Indian famine might have played out if the new high-yielding crop varieties had been available—but had been denied to struggling Indian farmers due to the enforcement of IP rules. The late Norman Borlaug, the father of high-yielding wheats, commented on this issue: "How will resource-poor farmers of the developing world, for example, be able to gain access to the products of biotechnology research? How long, and under what terms, should patents be granted for bio-engineered products? Further, the high cost of biotechnology research is leading to a rapid consolidation in the ownership of agricultural life science companies. Is this desirable? These issues are matters for serious consideration by national, regional and global governmental organizations."[17]

In view of these developments, a small but vocal group of scientists insists that improved varieties of basic food crops must continue to be delivered free or at minimal cost to the world's farmers, as public good or "open source" intellectual products to be shared by all humanity. Some even advocate the use of IP and patenting as a way of preventing the monopolization of food crops by large companies.[18] The gold rush by biotech and seed companies to patent plant genes that has taken place in recent years makes it likely, however, that a diminishing proportion of novel food crops will be freely available to farmers. This too, is a significant stumbling block to global food security.

It is important to understand that private R&D—which is worth about US$16 billion annually—has a huge amount to contribute to making global agriculture more efficient, especially in areas such as advanced crop breeding, smarter pest control, nutrient management, and precision farming technologies—just as it has in advanced medicine. These are the technologies that will be needed in the future to feed most of the world's population, especially in the great cities. There also remains a critical role, however, for publicly funded and philanthropically funded R&D in the large areas that private research does not fill. Ironically, the existence of a vibrant and growing private R&D sector is one of the reasons why many advanced countries have felt less need to devote taxpayers' money to agricultural research, cutting back on it on the venial pretext that the private sector was doing enough research to meet the need. This policy has exacerbated the gaps in public-good science, setting the world up for greater instability and risk of famine. There is also the plain fact that the sophisticated energy- and chemical-dependent farming systems now used in developed countries are a radical departure from the methods employed on smallholder farms in the third world; this means that the river of technology "spillovers" from the developed to the developing world is drying up—less knowledge is becoming available to poor farmers of a sort that they can easily use to raise their food output. These technologies are too costly, are unsuited to small farms, demand extensive training and high educational levels, depend heavily on scarce and expensive inputs based on fossil fuels, or else need to undergo prolonged and expensive adaptation for use in smallholder agriculture—which developing countries' national agricultural agencies rarely have the resources to carry out.

Another reason for the decline in emphasis on global food production is that, in advanced countries, more public research—loosely categorized in their budgets as "agricultural"—is in fact going to the food manufacturing sector. This research is usually defined as "value adding," meaning that its aim is to increase the price of a foodstuff to the consumer by adding some supposedly desirable feature to it. The issue here is that, when budgets are tight, public money spent on value-adding research cannot be spent on increasing food production on the farm. The ultimate effect of this may be to decrease global food security.

MAGIC BULLETS?

Depending on where its most vehement advocates and opponents sit, genetically modified (GM) crops are either the answer to the world's food problems or the work of the devil. No wonder the public is confused. The truth, as always, is more prosaic. By 2005, GM crops were reported to be grown by ten million farmers on a million square kilometers (250,000 square miles) of the world's farmland, and five trillion GM meals had been served—claims disputed by critics of GM crops.[19] The main GM crop–growing countries included Brazil, the United States, Argentina, Canada, South Africa, India, and China. Most of these crops have been endowed by scientists with extra genes to protect them against attacks by insects, diseases, or herbicides used to control weeds. Although these attributes may have lessened the use of insecticides on farms, leading to cleaner food and a healthier environment, evidence for the claim that they have increased global food output is thin. Although GM crops (such as cotton, soy, and canola) have in recent years made faster yield gains than non-GM crops (such as wheat, rice, and sorghum), it is not clear whether these gains are attributable to the actual transfer of new genes or are simply due to the fact that crops like cotton, soy, and rapeseed had more potential for yield gain, having previously been less intensively bred than wheat or rice.

Controversies over GM crops range from perceived risks to human health and the environment to ethical and religious objections, the potential impact of GM crops on the poor, control by large companies of the food supply, and the denial of choice to consumers. Indeed, it is fairly clear from the way the debate has evolved that, generally, the last people to be consulted about GM food were the people being asked to eat it and, having wide prior experience of technologies with undesirable side effects, citizens in many countries objected, resulting in a number of bans, restrictions, or consumer boycotts. In short, the story of GM is an object lesson in how not to introduce a powerful new technology—without at least consulting the people most affected and responding to their concerns. By not even attempting to do so, proponents of the technology succeeded only in retarding its uptake and gave their critics an easy target.

That said, biotechnology offers real promise for tackling some of the fundamental drivers of the coming famine—the shortages of water, land, nutrients, emerging diseases, soil problems, erratic climate, and the like. The world's food supply is not sufficiently secure that we can afford to turn our back on any technology that may help to address these in a safe and sustainable fashion, and especially one that can accelerate the delivery of new types of crops to meet urgent needs. The lesson of the early years of GM food is that modifications must be in tune with the wishes and needs of consumers at large and farmers in particular—not simply the genes preferred by scientists or agribusiness. These were the predominant focus of the first generation GM crops, but the second generation offers much wider scope for crops that are more tolerant of drought, flooding, heat, salinity, acidic soil, and climatic variability; more resistant to diseases or insect attacks; more able to extract nutrients from depleted soil or fix their own nitrogen from the air; higher yielding through hybridization and other enhancements; and better suited to local conditions. GM also offers the potential for novel crops endowed with qualities that will reduce the risk of cancer, heart disease, diabetes, and other ailments of modern society—which now affect countries both rich and poor.[20] Many of these qualities can also be delivered, though more slowly, through conventional plant breeding as well as GM. For the sake of food security as well as for reasons of consumer preference it will be important to pursue both approaches and not to neglect conventional breeding in favor of molecular methods of plant improvement or vice versa.

GM crops are sometimes depicted by their critics as being characteristic of a Westernized farming system that is anathema to smallholder agriculture. In reality, both sorts of agriculture can take advantage of the tools of biotechnology to reduce inputs of fuel, chemicals, and fertilizer and raise crop resistance to pests and diseases, as they need to do. In other words, biotechnology potentially holds benefits as great or even greater for the small subsistence producer and the organic farmer compared to the large modern commercial farm, as it may enable the small producer to sidestep the high-cost, high-intensity chemical farming route—yet still achieve greater outputs of food. Also, some agricultural challenges may simply not be solvable without the use of biotechnology.

One of the most formidable barriers to raising global food production is "harvest index." This is the ratio of the edible parts of a plant (such as its seeds) to the whole of the plant including its leaves, stem, and roots. That most of the world is well fed today is due to the remarkable achievements of plant breeders over the past hundred years in developing plants that convert more of the sunlight energy they gather into grain than into their inedible parts. There is a limit to how far you can push this ratio, however—plants have to have *some* leaves, roots, and stems—and some scientists are warning that, with our best new crop varieties, we are already close to it. Wild plants devote about a fifth of their photosynthetic energy to seed production. Today's domesticated crops devote up to half, and the theoretical limit is 60 percent.[21] In other words, the potential to redouble grain yields may be limited by what plants can physically do—and the hope that we can go on increasing crop yields for as long as we like may be dangerously misplaced.

In a quest for ways to get around this natural barrier, researchers at CIMMYT have delved into the ancestors of wheat, discovering that—unlike most of us—it actually had three parents, not two, an event that occurred through natural cross-pollination of wild grasses on the Anatolian uplands thousands of years ago. By imitating this process using natural plant breeding methods, the researchers have bred more than a thousand "synthetic" wheats, in which the original three ancestor grasses have been recrossed in different ways, to develop novel food crops capable of resisting disease or tolerating drought, heat, salt, or waterlogging, and lifting grain yields in some cases as much as 20–30 percent. In the process, breeders inevitably obtain strains with both desirable and undesirable genes, and CIMMYT's Tom Lumpkin argues that biotechnology will be needed to both hasten and broaden the process of developing new crop varieties by concentrating on the desirable genes and excluding the others.[22]

Another striking possibility is a project at the IRRI to reengineer the rice plant to capture sunlight and carbon dioxide more efficiently. Rice uses a photosynthetic process known as C3, whereas crops such as maize have a more efficient mechanism known as C4, which evolved from C3 plants around thirty-five million years ago. By searching among the wild ancestors of rice, the IRRI team hopes to discover one close to this ancestral divergence—and use it to breed a rice plant with a new cellular structure that enables it to use sunlight more efficiently, take up more carbon dioxide from the air, use less fertilizer and water, and yield more grain. If achieved, this would be among the most significant scientific advances in human history, changing what has been called "the most important chemical reaction on earth"—photosynthesis.[23] It is unlikely, however, that either supercharged rice or superwheats will be realities on most farms globally for at least a generation; indeed, it may be years before we know whether they are even capable of redoubling crop yields.

In any event, the world will need all the tools at its disposal to raise crop yields if it is to avoid the coming famine. Said the Green Revolution hero Norman Borlaug, "For the genetic improvement of food crops to continue at a pace sufficient to meet the needs of the 8.3 billion people projected in 2025, both conventional breeding and biotechnology methodologies will be needed."[24]

SOIL SOLUTIONS

While biotechnology has been the focus of agricultural science investment in recent years, in a time of slim budgets fields of equal promise and importance have suffered from comparative neglect. Among these is soil microbiology—the study of the billions of microscopic plants, microbes, fungi, and other life forms that live in the soil. A simple handful of dirt is said to contain a billion living organisms, many of which affect the growth of plants in both positive and negative ways. In the same way that farmers have learned to manipulate the soil surface to get more food from it, there is, in theory, potential to manage the life within the soil to obtain much higher yields of food.

Some organic and biodynamic farmers would claim to be doing this already—but they are doing it on the basis of judgment and belief, without a scientific understanding of the complexity of the "rainforest" of life within the soil and how these millions of microscopic bugs, fungi, and plantlets interact with one another, with crops, and with the surrounding soil. This is a task no less complex and challenging than understanding how the electrochemistry of the human brain leads to thoughts, feelings, and ideas, but for the purpose of avoiding the coming famine it is far more urgent as a scientific challenge. By understanding and managing the life within the soil we can perhaps

- unlock inaccessible nutrients within the soil, thus reducing the need for artificial fertilizers,
- control soil-borne pests and diseases that reduce crop yields by attacking the roots,
- stimulate greater plant growth,
- make thriftier use of scarce water,
- ameliorate hostile growing conditions, such as toxic, saline, or acidic soil,
- modify poor soil to make it better suited to food production,
- manage soil to store more carbon, so as to improve fertility and combat global warming, and
- discover new sources of edible protein among the tiny plants, fungi, and microbes in the soil.

The debate between "organic" and high-energy–based farming is a philosophical divide the world, in its present state, can ill afford. In one sense most of the world's farmers are organic—the smallholders who produce their own food and use few or no chemicals, fertilizers, machinery, or fossil fuel inputs. The farmers who grow the most food, however, are those who combine these technologies in ways that are increasingly sustainable and obtain more food from less land, water, and other inputs and with less damage to the soil. The time has come to put organic farming systems on a scientific footing—to understand how they work so as to get the very best out of them, and to bring together these two strands of agricultural thinking into a new form of "environmental agriculture" so each may benefit from the other. This strand of thought has been pioneered by respectable research establishments such as Rothamsted in the United Kingdom and the Rodale Institute in the United States over many years: it now needs to be extended to farming and grazing systems worldwide.

Research in this area is unquestionably vital for the world's 1.3 billion smallholder farmers, many of whom cannot afford the transition to advanced, high-intensity agriculture with its overwhelming reliance on fossil fuels, machines, and chemicals. But, facing peak oil, peak phosphorus, a shrinking water supply, limited land, and climate challenges, even high-intensity farmers now have an urgent need to find new low-input ways to grow food and can undoubtedly benefit from some of the ideas of the organic sector, provided they can be placed on a scientific footing. It is from the fusion of traditional science-based farming and philosophical approaches such as biological farming, permaculture, and organic agriculture that the new low-input, environmental farming systems of the twenty-first century will probably arise. This is not an area where the private sector has yet displayed great interest in investing, so it will probably depend largely on publicly funded science, which, as we have seen, is in a sadly run-down condition and has in any case tended to neglect research into low-input production in favor of that driven by large companies with inputs to sell.

Another area vitally requiring investment is in the handling, transportation, and distribution of food, both on the farm and after it leaves the farm, in order to reduce the colossal waste that currently occurs between farmer and consumer, especially in developing countries. In developed countries, too, far more research needs to go into solving the problem of waste of food. These are public-good objectives and are unlikely to be addressed comprehensively by the private sector alone. They demand a reinvigorated public/private research effort because—as we have seen—they lie at the heart of our ability to deal with the looming scarcities of land, water, energy, and nutrients that menace global food security.

NEW PLAGUES

Among the many challenges the world's agricultural scientists face in maintaining and improving the food supply, few are more urgent than keeping diseases of crops and livestock in check. In recent times, as humanity has cleared forests, intensified farm production, increased livestock numbers, and let down its guard, plagues never before seen have broken out, threatening both food production and sometimes human lives. Emerging threats such as nipah virus, chikungunya, and hendra virus are adding to the toll taken on the world's food supply by familiar pestilences such as foot-and-mouth disease, rinderpest, African swine fever, avian influenza, Newcastle disease, and Rift Valley fever. "Upsurges in animal disease emergencies worldwide are linked to the increased mobility of people, goods and livestock, changes in farming systems and climate, and the weakening of many livestock health services," says the Food and Agriculture Organization of the United Nations. "In both developed and developing countries, outbreaks have sometimes eluded the attention of central veterinary authorities for days or even months, allowing them to spread unchecked. The result has been unnecessary production losses, and growing difficulty in mounting effective control and disease eradication campaigns. These trends indicate that early warning is one of the weakest links in disease surveillance systems, at the national, regional and international levels."[25] The surge in crop and livestock diseases, as humanity intensifies its dominance of the Earth's biological systems, is another brake on increased global food production: it means that the scientists who should be working out ways to produce more food are tied down, fighting rearguard actions against new and emerging plagues. While human diseases such as SARS and bird flu receive immediate international attention and vast funding, efforts to contain the emerging diseases that menace global food security continue to starve in the dark. As the case of Ug99 wheat rust shows, agricultural diseases have the potential to play havoc with regional food supplies, creating large-scale hunger—and not only in the developing world.

The handful of examples given in this chapter help to illustrate the vast scale of both the challenges and the opportunities for food production, and the supreme importance of the scientific effort in meeting them. Although the eight most powerful men and woman on Earth, the leaders of the G8 countries, called for renewal in the world's agricultural science effort, it is doubtful that they really appreciated how far the knowledge machine has been allowed to run down, how long it will take to restart it—or how vast and complex is the challenge that lies ahead in attempting to stave off the coming famine.

Yet it must be said that, if water, land, nutrients, energy, and stable climates are all increasingly scarce, the one thing not in short supply is brains. It is high time we used them more: now is the moment when *Homo* gets to earn the tag *sapiens*.

SHARING KNOWLEDGE

It is often said that the world has plenty of food—it is just very poorly distributed. The same is true, and more so, of agricultural knowledge. There is a vital difference, however. Poor people who have no food cannot afford to buy it from rich nations who grow it using costly methods. And even if governments agreed to drop all trade barriers to the movement of food worldwide, there would still be the vast logistical challenge of delivering it to places where port, rail, road, and refrigeration services are poor or nonexistent, so much of the food would rot before it reached those who need it. Also, giving food away is no permanent solution, as it undermines the ability of people to support themselves, creating a culture of dependency and depriving them of the main launchpad for economic growth and self-governance. The best solution is nearly always for local people to be able to grow as much of their own food as possible—and to have the right to choose their own ways of doing so, in conformity with their climate, soil, water, culture, and beliefs.

In the past this difficulty has been addressed mainly in two ways—by delivering seeds and by delivering knowledge. Seeds are, in fact, little packages of knowledge, specially designed to perform in particular environments and conditions. They are highly portable and easy to deliver both to smallholders and to advanced farmers. Often, however, they require specialist skills and expensive inputs such as chemicals or fertilizers to get the best out of them—and this applies to all high-intensity farming techniques.

Knowledge, on the other hand, is delivered by many and various routes to farmers, both well-off and poor—by scientists and agricultural extension workers, by agribusiness companies, by farming media and the Internet, by nongovernmental organizations and aid agencies, by fellow farmers, and even by agents who buy farm produce. Compared even with the depleted scientific enterprise, the world agricultural knowledge chain is ramshackle, far more neglected and only as strong as its weakest link. One of the main reasons that advanced farming methods have caught on in some countries but not others is that the local agriculture departments have lacked the resources and skills to deliver the knowledge to millions of smallholders. I recently spoke with a highly skilled officer in an African agriculture department who, although he had knowledge of the latest methods, keen staff, and a four-wheel-drive vehicle, had no paper on which to print information for farmers and no fuel for his vehicle to take the knowledge out to field days or grower meetings. This illustrates how the best agricultural science in the world can fail to achieve its promise if it is poorly delivered or not delivered at all.

It is sometimes said that if the existing knowledge we have about food production today were better distributed to farmers so they could use it, we could redouble global food output—and there is undoubtedly an element of truth in this. Without going into detail, massively more can be done than we do at present to deliver knowledge to farmers, large and small, well-off and poor, especially using modern, low-cost, and innovative mass communication methods—and if this were achieved it would have a more beneficial impact on world food security than the biggest scientific breakthrough or greatest technical innovation.[26] One of the vices of the present global R&D system is that it values, and invests in, knowledge creation much more highly than knowledge sharing. As a result, the communication of knowledge with farmers continues to lag far behind, and if the world is serious about solving the food crisis it will need to match every scientific research dollar with a dollar to deliver that knowledge to farmers and consumers. There is hardly anything as regrettable as a fine scientific advance that is never put to use. The creators of mobile phones, music players, and expensive pharmaceuticals would never dream of producing a valuable new product and not telling their customers about it; in agriculture it happens all the time.

WHAT CAN I DO ABOUT IT?

Insist that our governments invest more in sustainable agricultural research and development, in international agricultural research, and in sharing the resulting knowledge.

EIGHT

EATING OIL

We can evade reality, but we cannot evade the consequences of evading reality.

—Ayn Rand

Most people haven't a clue how much oil they eat.

For a person on a typical Western diet, one estimate is around 4.4 liters (about 1.2 U.S. gallons) of diesel a day, meaning that it takes the distillate from 66 barrels of crude oil just to put their food on the table for a year.[1] A well-off family consisting of two parents and two children "eats" 175 barrels of oil—almost one barrel every two days. Even in the developing world, families on rising incomes and city inhabitants now depend largely on oil-based food supplies.

Most of the food on our plates has been grown, transported, processed, chilled, and stored using vast quantities of fossil energy. According to the ecologists Mario Giampetro and David Pimentel, it takes about ten calories of fossil oil energy to produce a single calorie of food energy.[2] The global food miracle of the past half century would not have happened without the oil that powers the tractors, manufactures the fertilizers and pesticides, trucks things on and off the farm, and ships them around the world. Humanity would be far, far less numerous today were it not for the massive infusion of energy we received in the second half of the twentieth century from oil and the subsequent boost in fertility and reduction in death rates it imparted. The French chef Brillat-Savarin remarked that we are what we eat: in school they teach us that humans are largely made from water, but in a sense most of us are mainly made from oil.

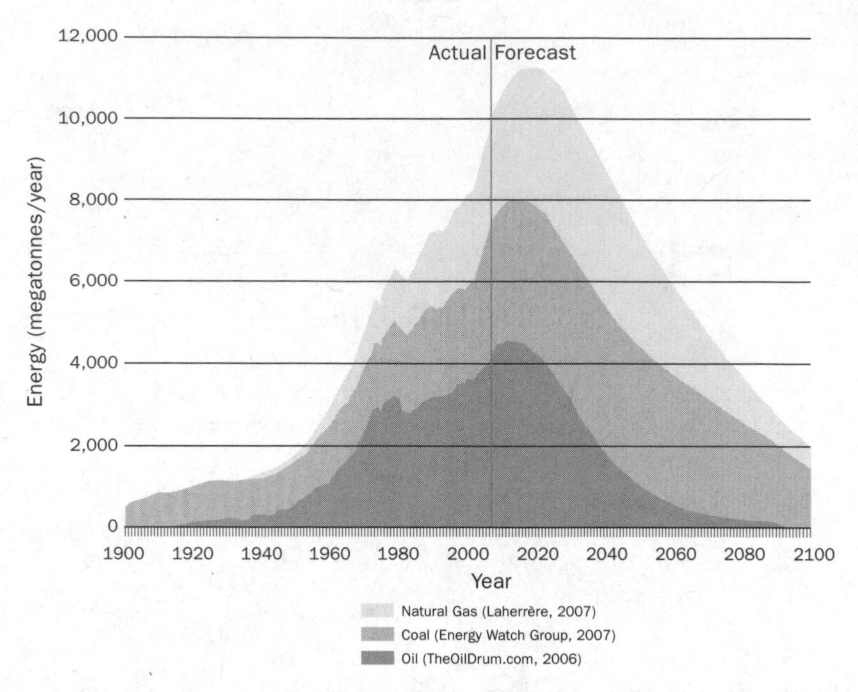

Figure 6 Production of fossil fuels, 1900–2100. Source: Jeremy Gilbert, "How Will We Eat When Oil Runs Low?" slide 4 in presentation at Fuelling Food in Western Australia conference at University of Western Australia, Perth, October 3, 2008.

A second important issue is that few, if any, of this generation's grandchildren will be "eating oil." The use of fossil petroleum to power the world's food-production system will probably be long gone by 2050 and, unless we've found cheap, reliable substitutes in the meantime, then so will much of the civilization that depended on it (figure 6).

Peak oil—the idea that sooner or later world demand for oil will overtake supply—is no longer a mystery, in spite of the fact that many governments around the world avoid discussing it and frequently behave as if it weren't happening.[3] The reason is that no really huge oil discoveries have been made anywhere in the world for a half century or more; those that were made back then are starting to run low, while demand for oil both in the West but also especially in China and India has been growing at about three times the rate at which new discoveries are being made. Domestic oil production in countries such as the United States, Britain, and Australia is already well past its peak. This growing uncertainty prompted a spectacular hike in world oil prices in 2008, followed by an almost equally spectacular collapse: extreme price volatility is a sign that supplies are becoming far more uncertain. Also, oil price volatility and supply uncertainty are extremely dangerous for food production.

Colin Campbell and Jeremy Gilbert—former chief petroleum engineer with British Petroleum—lucidly explain what your leaders will not tell you: "Oil was formed in the geological past: we cannot grow more. People ask: 'Are we running out of oil?' The simple answer is: *Yes, we started doing that when we produced the first barrel.* But finally running out is not the issue. That won't happen for a very long time, if ever. Much more important is the pattern of depletion. Production starts to decline when about half has been consumed. Decline is primarily driven by the immutable physics of the reservoir, although political and economic factors may intrude. This is the issue that should concern us. It is a very serious one."[4]

Campbell and Gilbert say that every beer drinker should be able to grasp the idea. "The glass starts full and ends empty. The quicker you drink it, the sooner it is gone. The bar shuts at closing time. The same principle applies to oil and gas. We may ask how this self-evident reality has been concealed and confused. It is a devastating realization with huge implications for the modern world. We may lack the mental courage to accept it, but Nature does not lie." Campbell and Gilbert point out that the ways oil companies and oil-producing nations report their estimates of reserves are often designed to mislead and confuse, usually with a view to manipulating the value of their shares or product in the market. As a result of this universal habit of deception, nobody really knows when peak oil was reached, or whether it still lies just ahead of us. In Campbell and Gilbert's view, it is very close to now, and this means that oil and the products made from and with it can only become more expensive over the long haul, even though global recession may for a time mask the trend by dampening demand and prices. The International Energy Agency, a conservative body, has warned that peak oil will be reached globally between 2010 and 2020.[5]

Currently the world uses about 11 billion tonnes (12 billion U.S. tons) of primary energy a year to extract, produce, and transport 100 billion tonnes (110 billion U.S. tons) of materials, among which by far the most important is food. If primary energy declines, then so does production and so do economies . . . or so the prevailing economic theory says. On the view that the market will fix this problem, Pedro Prieto, vice president of AEREN, the Association for the Study of Energy Resources, and the Spanish member of the Association for the Study of Peak Oil, is scathing: "Classical economists still work, think and behave as if they lived on a flat Earth with unlimited resources to be made available by Man's ingenuity and market forces. Very serious issues are at stake caused by the growing gap between the available fuel supply, which is subject to natural depletion[,] and the ever-growing demand implied by classical economic theory."[6] The effects of tightening supplies were evident in the oil price surge of 2008—airlines charging more, farmers using less fertilizer, car sales declining and consumers seeking more fuel-efficient vehicles, and power shortages in some countries. Hardest hit in food terms were farmers in emerging and developing countries that had just begun the switch to the Western, oil-intensive system of food production to feed their mighty cities. China has increased its fertilizer usage 44 percent in recent years, India 33 percent, Pakistan 61 percent, and Brazil 137 percent. Europe, by contrast, has cut its fertilizer use by half.[7]

The eco-commentator Richard Heinberg identifies four interlocked dilemmas that oil volatility now presents to humanity: "The first dilemma consists of the direct impacts on agriculture of higher oil prices: increased costs for tractor fuel, agricultural chemicals, and the transport of farm inputs and outputs," he says in his eloquent essay "What Will We Eat as the Oil Runs Out?"

> The second is an indirect consequence of high oil prices—the *increased demand for biofuels,* which is resulting in farmland being turned from food production to fuel production, thus making food more costly. The third dilemma consists of the impacts of *climate change and extreme weather events* caused by fuel-based greenhouse gas emissions. Climate change is the greatest environmental crisis of our time; however, fossil fuel depletion complicates the situation enormously, and if we fail to address either problem properly the consequences will be dire. Finally comes the *degradation or loss of basic natural resources* (principally, topsoil and fresh water supplies) as a result of high rates, and unsustainable methods, of production stimulated by decades of cheap energy.[8]

The issue of energy for agriculture is very similar to the issue of virtual water, discussed in chapter 3. In theory, countries that lack water can import "virtual water" as food commodities from those with plenty. So too, countries that lack the energy to grow all their own food can import surplus food from countries with highly productive oil-based farming systems—provided they are rich enough to afford it. The fact, however, that a billion people starve while another billion wallow in surpluses of food so huge that they throw away half undermines this idea. As price and supply uncertainty bite home over the coming generation, the cost of food produced by such systems will soar in sympathy, battering not only the poorest countries but also the poor and middle classes of richer countries and creating a recurrent source of instability sufficient, perhaps, to conjure up specters of the famines that brought down the monarchies of France and Russia. Based on current oil supply projections, by 2050—when food demand is forecast to be double what it is today, to feed nine billion people—there will be about as much oil available as there was in 1960, when there were just three billion people on Earth, a third of whom were starving. Whatever powers the world food-production system then, it mostly won't be fossil oil.

Warnings about the global food system's overreliance on an energy source that will *inevitably* become inaccessible through both price and supply are not news. In 1994, Henry Kindall and David Pimentel noted it as one of the greatest constraints to expanding the global food supply, requiring an amount of fossil energy almost equivalent to all the solar energy used to grow the world's crops.[9] Just when the crunch point will come, no one has been able to predict with certainty—but the more world living standards rise, the faster it will approach. Alarmingly, many of the so-called solutions to the global food crisis—high-yielding crops, highly automated precision farming systems, and methods of reducing losses of food both on and off the farm (e.g., refrigeration, drying) are largely dependent on fossil energy. So too is the entire world trade in food.

Yet, says Swedish physicist Kjell Aleklett of Uppsala University's Hydrocarbon Depletion Study Group, "The International Energy Agency (IEA) estimates that, during the coming twenty years, approximately 115 billion barrels will be located, an estimate that we agree with. During the same twenty-year period they also forecast that the world will consume more than 600 billion barrels of oil."[10]

At some dim level these issues have begun to register with governments but their general answer has problems enough of its own and may, in many instances, not be an answer at all: biofuels.

GOOD FUELS, BAD FUELS

When global food prices soared in 2007–8, biofuels were swiftly transformed from greenhouse hero to food price villain in the eyes of the public and a good many politicians and commentators. Up to that time they had been broadly viewed as benign, one of the possible solutions to imported oil dependency, climate change, and rural economic development. A few commentators had warned that they could potentially compete with food—but that was before prices took off. Then the World Bank economist Donald Mitchell penned a note in which he attributed three-quarters of the spike in food prices to the expansion in biofuels production in the United States and the European Community—and the consequent effects of this on grain stocks, land use, speculation, and trade. The assessment sparked a row with the Bush administration, which had attributed only 3 percent of the rise in food prices to biofuels, and it focused the glare of public attention on an industry that most people had regarded, until then, as fairly insignificant in the overall scheme of things.[11]

Of the world's 2006 fuel ethanol production of around 40 billion liters (10 billion U.S. gallons), nine-tenths was produced in Brazil and the United States, and of 6 billion liters of biodiesel (1.5 billion U.S. gallons), three-quarters was produced in the European Union, chiefly in France and Germany, another World Bank study reported. It said,

> Brazil is the most competitive producer and has the longest history of ethanol production (dating back to the 1930s), using about half its sugarcane to produce ethanol and mandating its consumption. With tax incentives, subsidies, and consumption mandates for biofuel production, the United States used 20 percent of its maize crop to produce ethanol in 2006/07. New players are emerging. Many developing countries are launching biofuel programs based on agricultural feedstocks: biodiesel from palm oil in Indonesia and Malaysia, ethanol from sugarcane in Mozambique and several Central American countries, and ethanol from sugarcane and biodiesel from oil-rich plants such as jatropha, pongamia and other feedstocks in India. Current biofuels policies could, according to some estimates, lead to a fivefold increase of the share of biofuels in global transport energy consumption—from just over 1 percent today to around 5 to 6 percent by 2020.[12]

Most current biofuels are not economically viable without subsidies: indeed, demand for them would collapse if they were priced at their real cost of production. According to various estimates, the United States props up its domestic ethanol industry with more than two hundred measures collectively costing taxpayers from six to twelve billion dollars a year, equivalent to a subsidy of thirty to one hundred cents on every liter sold.[13] The cost of producing biofuels depends mainly on two things—the price of oil and the price of the grain feedstock. Unfortunately, as the price of oil rises, so too does the price of agricultural commodities, which in turn pushes up the price of the biofuel feedstocks. The problem cuts both ways, as the production of farm-based biofuels also drives up the cost of food. The International Food Policy Research Institute has calculated that if global biofuels production doubles in its current form, by 2020 it will push corn prices 26–72 percent higher than they would otherwise be, oilseeds 18–44 percent higher, wheat prices 8–20 percent higher, and sugar 12–27 percent higher—hitting all consumers in the shopping basket.[14]

Biofuels thus pose a range of challenges:

- They may compete with food crops for scarce land and water.
- They compete with food for nutrients, energy, and pesticides—all fossil-derived products. The Food and Agriculture Organization of the United Nations (FAO) cites estimates that fuel crops could consume between 2 and 14 million tonnes of fertilizer per year.[15]
- If grown on marginal land, they may worsen soil loss and greenhouse emissions.
- "First-generation" biofuels such as ethanol from grain take almost as much energy (including significant amounts of fossil energy) to produce as they yield.
- They raise ethical issues: the amount of grain needed to fill the tank of a sport utility vehicle just once would feed a poor person for a year, the World Bank notes. To grow the grain to make one tankful of diesel fuel requires 300 tonnes (330 U.S. tons) of water.[16]

Were recent trends in biofuels production to be maintained, by the 2020s we would be burning the grain equivalent of the total world rice harvest just to keep cars and trucks on the roads. Many countries now acknowledge that this is unacceptable and are modifying their policies. Scientists have also shown that biofuels are not as good for the climate as their supporters have made out: if nitrogen fertilizer is used to grow them, it releases a greenhouse gas (nitrous oxide) three hundred times more potent than carbon dioxide, and other emissions occur during the production process. If new land is opened up to make way for fuel crops—as is happening in parts of Asia—every hectare cleared releases hundreds of tons of carbon into the atmosphere and it may take anywhere from seventeen to four hundred years for the fuel crop to compensate for this release, depending on the type of land cleared and type of crop grown.[17] This implies that biofuels are likely to be an immediate net benefit to the climate only if grown on existing cropland—where they compete with food. In fact, if the aim is to reduce greenhouse emissions, it is more effective to turn cropland back into forest than to grow biofuels on it.

That said, biofuels offer farmers a degree of insulation from the vagaries of the world oil supply. For example, John Hamparsum, a leading Australian grain and cotton farmer, calculates that, in the event of a petroleum shortage, he would need to set aside about one-tenth of his land to grow canola for biodiesel to supply all his on-farm fuel and electricity needs.[18] (This compares interestingly with the average horse-powered Western farm of a century ago, which had to use about one-fifth of its land to grow feed for the draft animals.) Although the proportion of land needed will vary, the implication is plain: if world agriculture of 2050 were forced to grow all its own fuel, this could reduce the global food supply on the order of one-tenth or more—at the very time it needs to double. If farms needed to grow sufficient fuel to transport, store, and distribute their produce and inputs, then global food production might diminish by as much as one-third. So, although the farm sector is perfectly capable of meeting its own energy needs, agricultural energy-sufficiency comes at a cost of producing significantly less food at a time of growing scarcity.

The second-generation biofuels are less likely to compete directly for food-producing land and water. These will be made from crop and timber wastes, scrub collected from rangelands, and also specialist energy crops such as giant napier grass or giant reed grown on land that cannot be used to grow food, such as polluted mine sites and municipal garbage dumps. These lignocelluloses can be converted into ethanol or biodiesel using a range of enzyme-based or thermochemical processes.[19] Cheap, efficient technologies to convert their cellulose into sugar for ethanol production or to gasify their biomass are not yet commercially viable, however, and may take some years to arrive. Also, energy crops still need fertilizer and possibly pesticides, and thus compete indirectly with food crops. The International Energy Agency estimates that to satisfy 10 percent of the transportation needs of the United States or Europe with biofuels by 2020 would occupy up to 30 percent of their total cropland area.[20] Thus, on their own, biofuels are quite unlikely to provide a lasting solution to the world's transportation problems.

There are, however, many clever ideas that involve farm energy. In the Philippines, for example, the agricultural engineer Alexis Belonio has invented what could prove a lifesaver for millions. He started by building a simple cooker that converted rice hulls—the waste left after rice is husked—into a gas that burns with a hot, blue flame. The world discards 150 million tonnes of rice husks every year, and Belonio's US$20 stove not only provides a valuable use for the waste but could also help to address one of the major drivers of global warming: black carbon. Black carbon is the soot produced by billions of cooking fires and household stoves in the developing world. Scientists estimate it contributes half as much to warming the planet as the total carbon dioxide emissions from fossil fuels. However, because black carbon stays in the atmosphere for only about ten years (whereas carbon dioxide stays up there for centuries), removing it could rapidly have a large impact on global warming. This requires global distribution of Belonio stoves to families who currently use open fires, wood-fueled ovens, inefficient biomass stoves, or cookers fueled by kerosene or gas. The char left by these gasifiers can be used to enrich farm soil. Belonio says the cookers can also burn sugarcane bagasse, soybean hulls, cacao husks, and other crop wastes. Alexis Belonio gives his technology away free to benefit as many people as possible, and for this extraordinary generosity he was acclaimed an associate laureate of the Rolex Awards for Enterprise.[21]

Another dimension of the new-style biofuels is illustrated by a British company, D1 Oils, which plans to plant a million hectares of an oil-rich plant called jatropha on marginal land in India, Indonesia, and Zambia. These, it claims, will produce about 12 million barrels of oil a year, rising to 19 million when new varieties of the crop are introduced. Assuming the company can restrict plantings to nonfood land—doubtful if the crop proves profitable—this is still only about a fifth of what the fossil oil industry produces in a day, though it may keep local farms running.[22]

A third dimension, and perhaps the most promising future source of biofuels, is algae farming. Humans have already demonstrated that we are incredibly good at growing algae where algae are not wanted, by spilling our wasted nutrients into rivers, lakes, estuaries, bays, and seas. These tiny water plants can produce oil for biodiesel and carbohydrates for ethanol, fertilizers, food for us, feed for livestock, fine chemicals, and pharmaceuticals such as beta carotene and omega-3 oils—and indeed are said to produce more oil per unit of plant matter than any land-based plants, around 2.7 tonnes (3 U.S. tons) per hectare.[23] Algae have aroused the interest of major airlines such as Virgin and aircraft corporations such as Boeing as a prospective green fuel source. A U.S. Department of Energy estimate suggested that America could meet its entire liquid fuel needs from an area of oil-producing algae slightly larger than Maryland. Algae grow in a wide range of environments, including those that are very saline, hot, and polluted.[24] Large uncertainties remain, however, including the cost of obtaining the oil, the right species of algae, the yields, the most efficient farming systems, the need to destroy vast areas of coastal marshes, wetlands, and salt lakes to build algae farms, and the availability of nutrients such as phosphorus to sustain what could undoubtedly become a gigantic new industry. A final major uncertainty is the availability of suitable land or coastline at a time of likely sea-level rise, or the availability of sufficient water to grow algae inland. Although algae may not be able to substitute for fossil petroleum entirely, they are nevertheless an important potential energy source to support global and local food production as well as a promising new industry creating jobs and economic activity in poorer places.

Other alternatives for on-farm energy include the use of biogas, made from fermented animal or crop wastes, for fueling tractors, heating sheds, or running equipment.[25] Vehicles and equipment powered by natural gas and/or fuel cells may be restricted by the extent of world reserves of natural gas, which are believed to be as limited as those of oil, and it is possible that liquid fuels made from oil shale, tar sand, and black or brown coal will prove too hazardous to the world climate to produce on a large scale. On smaller farms electricity from renewable sources such as sun and wind may provide significant motive, static, and transportation energy in the future, while for large-scale broadacre operations, hydrogen is perhaps the ultimate fuel once the difficulties of moving and storing this ultralight and fugitive gas are overcome. Mercedes Benz, which is now working on its fourth-generation hydrogen-powered vehicle, expects to have a commercial version on the road within a few years, and the benefits of this technology will undoubtedly spread to agriculture eventually.

On balance, however, it must be said that the world food system as a whole appears largely unprepared for the energy shortages and price spikes that lie ahead, and few governments have taken the trouble to seriously address Heinberg's question, "What will we eat as the oil runs out?"

THE LOW-ENERGY ROUTE

Another approach to the looming energy crisis in food production is to maintain or develop as much of world agriculture as possible as low-energy, low-input farming—using a minimum of fuels, fertilizers, synthetic chemicals, and transportation and relying instead more heavily on human brains and muscle.

A study by Jodi Ziesemer for the FAO concludes that organic farming typically uses 30–50 percent less energy than "conventional" energy-intensive agriculture. Similar conclusions have been drawn by other authors about biodynamic and permacultural systems. (It should be noted, however, that organic agriculture in developed countries can use large amounts of fossil fuel for plowing and weed control, which other farmers carry out using chemicals and conservation tillage.) Ziesemer's study also pointed out that food imports are often more energy-costly than food grown locally, and there is therefore much sense in developing local supplies close to or in big cities.[26] The debate over "food miles" has helped to clarify many of these issues, leading to a conclusion that locally grown foods are usually best, provided that local transportation and distribution chains are also energy- and cost-efficient. Where these are poor, there may well be both energy and cost advantages in shipping food from afar—an example being China's long reliance on imported grain to feed its coastal cities, rather than moving the necessary tons of domestic grain on constricted internal rail and road systems.

The case for world agriculture as a whole to go "local and organic" has been eloquently put by a number of authors such as Raj Patel, in his book *Stuffed and Starved: The Hidden Battle for the World Food System*. In a recent article, Patel argues for

> a deep shift in the way our food comes to us. More and more scientists are encouraging us to abandon the food system of the past century, and to go local and organic. Instead of industrial agriculture, they recommend increasing support for agro-ecological farming—a way of growing food that builds, rather than destroys ecosystems. Instead of spraying chemicals to get rid of pests, grow plants that attract beneficial insects. Instead of applying fossil-fuel-based fertilizers to the soil, a technique that destroys the soil's own capacity to regenerate, lace the fields with legumes, which naturally help to fix nitrogen in the soil.

"Improved farming science alone won't fix things, though," Patel continues. "As much as they need nitrogen in the soil, tomorrow's food systems need democracy on the ground. The problem of starvation is one not of production—we produce more than enough food to feed everyone—so much as poverty and distribution. To fix this deeper problem, progressive groups and citizens are showing national governments that the best way to solve hunger is through active citizen participation."[27]

Patel is one of an increasing number of voices being raised in favor of low-input smallholder agriculture, not only for the developing world but also for the developed, as it comes to grips with issues such as energy shortages, obesity, degenerative disease, and climate change. As we saw in chapter 4, smallholdings often grow more food per unit of land area than large farms do. They use far fewer chemicals and fertilizers and more human labor. They are also more diversified, growing many small crops, which is an advantage in terms of agro-ecology—but conflicts with the demands of the supermarkets and food companies for large lines of uniform produce, adds cost to the marketing chain, and increases postharvest losses. Countries such as Vietnam have achieved food self-sufficiency, however, by improving smallholder agriculture rather than moving to broadacre or corporate farming systems. The International Bank for Reconstruction and Development and the World Bank, in their *World Development Report 2008,* comment, with particular reference to Africa, that "using agriculture as the basis for economic growth in the agriculture-based countries requires a productivity revolution in smallholder farming."[28]

IDEOLOGICAL FARMING

Patel is right about something else, too—there is an ideological battle going on between the two styles of agriculture and their adherents, in which each is seeking to overwhelm the other. A recent manifestation of this is La Via Campesina (LVC), an international organization of more than one hundred peasant and smallholder bodies claiming to represent two hundred million people worldwide, which was founded in 1993 and is today headquartered in Indonesia. Asserting that it is independent of any political, economic, or other affiliation, the group upholds what it terms "food sovereignty," which it defines as "the right of peoples to healthy and culturally appropriate food produced through ecologically sound and sustainable methods, and their right to define their own food and agriculture systems." LVC says that it places the aspirations and needs of those who produce, distribute, and consume food at the heart of food systems and policies—rather than the demands of markets and corporations. It seeks to build a coalition of producers and consumers against landlords, corporate agribusiness, and government control of food.[29] Despite the contemporary international context, there is a strong flavor of agrarianism about LVC's ideals, expressed in the declaration of its second youth assembly:

> The countryside is our life
> The earth feeds us
> The rivers run in our blood
> We are the youth of the Via Campesina
> Today we declare the beginning of a new world
> We come from the four corners of the world
> To stand together in the spirit of resistance

> To work together to create hope
> To talk together about our struggles
> To learn from each others work
> To be inspired by each others songs, music and stories
> To build solidarity between our movements
> To unite as a strong force for social change.[30]

The ideological rift between the two schools of thought erupted into full view in 2007, when four hundred leading agricultural scientists reported to the World Bank on the knowledge needed to sustain the global food supply.

> The International Assessment of Agricultural Knowledge, Science and Technology for Development (IAASTD) responds to the widespread realization that despite significant scientific and technological achievements in our ability to increase agricultural productivity, we have been less attentive to some of the unintended social and environmental consequences of our achievements. We are now in a good position to reflect on these consequences and to outline various policy options to meet the challenges ahead, perhaps best characterized as the need for food and livelihood security under increasingly constrained environmental conditions from within and outside the realm of agriculture and globalized economic systems.[31]

The IAASTD study provided a timely reminder that farmers are the most important people in the world. They not only feed us but also manage most of our landscape, its water, and wildlife as well as a significant part of the atmosphere via soil carbon and emissions. The report warned of the dangers of "a world of asymmetric development, unsustainable natural resource use, and continued rural and urban poverty" and called in particular for the adoption of farming systems that are both sustainable and socially equitable and that give struggling farmers greater support not only to grow food but also to care for the world's resources of land, water, and wildlife. The IAASTD group also urged stronger public influence and control over biotechnologies such as genetic modification (GM) and was critical of the impact of current intellectual property policies on poor farmers.

Fifty-seven countries signed the report and three refused: the United States, Canada, and Australia. These countries did not state explicitly why, merely saying they could not agree with everything in the report; however, the report's stance on biotechnologies and intellectual property rights is known to have offended them. That three countries with such distinguished records of providing overseas aid should turn their back on worthwhile proposals to assist the world's poor and its farmers and to sustain the global food supply in this fashion is a sign of how deep the ideological rift is.

The philosophical divide between adherents of what we might call LISA (low-input smallholder agriculture) and HEFS (high-energy farming systems) will not feed the human race through the greatest surge in demand and chronic scarcity of resources in history. Ideologies alone never provide bread, though they frequently interfere with its production. At this juncture, arguing over which is the best system may prove lethal to millions upon millions of people, children especially.

The simple fact is that we will need both styles of agriculture working at their very best to have a chance of feeding humanity through this critical phase—and we need to take the best ideas from each and cross-fertilize them. Both farming systems have advantages and disadvantages and involve trade-offs that affect the farming environment in different ways: the goal must be to minimize these trade-offs and share ideas between the two systems. Just as diversity provides a buffer against external shocks in natural ecosystems, it creates resilience in farming systems too. Urban Americans, Europeans, and Asians are unlikely to abandon their supermarkets in favor of local organic farmers' markets—though there may well be a drift of individual consumers that way—and supermarkets will continue to rely on high-energy agriculture to a large degree for their basic foodstuffs while also catering to growing customer demand for organic products. Equally, if millions of poor farmers were to adopt Western high-energy farming systems, not only would they help burn up the world's oil and fertilizer faster than ever but the process would throw up to a billion of their fellow farmers off the land, massively augmenting the ranks of the urban poor—an outcome that has happened throughout history whenever some farmers adopt radically more competitive means of production.

Tugged at and battered by both sides in this ideological scrap, the international agricultural research system is being torn two ways. On the one hand, it appears to buy the argument that the path out of poverty for many poor farmers is to mechanize, use more fertilizers, chemicals, and high-yielding or GM crops, and enter the global food trade. On the other hand, it recognizes that this will inevitably leave a billion or so poor farmers, chiefly women, behind—and more must be done to help and sustain them within their local farming systems.

World agricultural science at the national level has for two generations been dominated by the high-energy model of farming—and it has produced a food supply miracle, which is now seen to have high social and environmental costs as well as great dependence on failing resources. Science has largely neglected (or not been funded to properly research) the equally promising but far less understood low-input systems or methods such as organics and permaculture. There can be little doubt that, as science focuses more intently on these complex systems, it will unlock new insights capable of making profound gains in food production and sustainability on a par with those of the Green Revolution, and in both styles of farming. This, perhaps, is where the future "eco-agriculture" is to be found.

The eminent Australian land and water scientist John Williams argues that devising farming systems that do not ultimately destroy their own environment and resources is perhaps the greatest challenge ever faced in the ten thousand years since agriculture began. "We need a whole systems approach—one that doesn't drink rivers dry, which recycles its nutrients, which does not impact on the wider environment," he explains. "To develop an agriculture like this will be one of the hardest things we have ever undertaken."[32]

In many countries the influence of corporate agribusiness over government agricultural laboratories and the prevailing view that high-energy farming systems offer the best—indeed the only—solution have hitherto combined to stall this research. This is *not* to argue that we should be doing less HEFS research; rather it is to say that we should be doing a great deal more research on both types of agriculture and should make every effort to combine the two streams of thought productively in order to create a new eco-agricultural system. High-energy farming systems may be at risk as the emerging energy, land, water, and fertilizer shortages of the coming quarter century reach their peaks—but an even greater danger is that squabbles over which is the better system will stall development of the new eco-agriculture.

The eco-commentator Richard Heinberg writes,

> Given the fact that fossil fuels are limited in quantity and that we are already in view of the global oil production peak, we *must* turn to a food system that is less fuel-reliant, even if the process is problematic in many ways. Of course, the process will take time; it is a journey that will take place over decades. Nevertheless, it must begin soon, and it must begin with a comprehensive plan. The transition to a fossil-fuel-free food system does not constitute a distant utopian proposal. It is an unavoidable, immediate, and immense challenge that will call for unprecedented levels of creativity at all levels of society.
>
> A hundred years from now, everyone will be eating what we today would define as organic food, whether or not we act. But what we do now

will determine how many will be eating, what state of health will be enjoyed by those future generations, and whether they will live in a ruined cinder of a world, or one that is in the process of being renewed and replenished.[33]

HOW MUCH ENERGY DO I EAT?

The British physicist David MacKay has calculated that the average person on a Western diet consumes 12,000 watt-hours of energy every day—which is the equivalent of what it takes to run 120 hundred-watt light bulbs for an hour.[34] Some of this energy is in the form of free sunlight, falling on the crops that form part of our diet or supply feed to animals. The rest is mostly fossil energy used to grow and transport the food.

A moderately active person with a body weight of 65 kilograms (143 pounds) consumes food with a chemical energy content of about 2,600 kilocalories per day, which is about 3 kilowatt-hours (3,000 watt-hours) per day. Most of this energy escapes from the body as heat, he says. Drinking a pint of milk and eating 50 grams (approximately 1.75 ounces) of cheese per day consumes 1.5 kilowatt-hours per day. Eating two eggs consumes a kilowatt-hour. Eating a half pound of meat each day (the average meat consumption of Americans) consumes 8 kilowatt-hours per day. MacKay calculates the energy content of typical diets as follows.

Standard diet of meat, dairy, vegetables, and grains	12 kilowatt-hours/day
Vegetarian diet (mostly vegetables, some dairy or eggs)	4 kilowatt-hours/day
Vegan diet (no animal products)	3 kilowatt-hours/day

MacKay also points out the incredible amount of energy consumed by pets every day:

Cat	2 kilowatt-hours
Dog	9 kilowatt-hours
Horse	17 kilowatt-hours

NINE

THE CLIMATE HAMMER

> Our agricultural systems have been adapted to climates which are about to become extinct.
>
> **—Cary Fowler**

Jutting like the prow of a phantom ocean liner forging fatefully into the eerie Arctic mists, the concrete doorway to the "Doomsday Vault" is emblematic of the perilous waters into which the human voyage is taking us. Opened in 2008, the Svalbard Global Seed Vault, on the remote Norwegian island of Spitsbergen, barely a thousand kilometers from the North Pole, is intended to rescue the world's food system from catastrophe by preserving the seeds of our essential crops so they can be multiplied for resowing in the wake of war, famine, fire, flood, or storm. Within its frozen walls, tunneled 125 meters (410 feet) into the bleak permafrost and mountainside of this Arctic isle, a collection that already harbors more than a hundred million seeds is being assembled. They range from staples such as maize, rice, wheat, lentils, and sorghum to eggplant, tomato, lettuce, barley, banana, and potato.[1]

The skipper of this landlocked vessel, the crop scientist Cary Fowler, speaks of the gathering of a "perfect storm" of events that menace the world's food supplies, referring to the dwindling of the world's rivers and groundwater, the vast demands for increasingly scarce energy, the growth in human need for food and the shrinking of grain reserves, the underinvestment in science, and, above all, the devastating changes in climate, which are already starting to be felt—but with the worst yet to come. "In historical terms, this means that the hottest growing seasons of the past hundred years or so would in the future become the coldest

growing seasons," Fowler cautions. "Present-day agricultural systems and crops have been adapted to specific regions with particular climates since Neolithic times. Many crop varieties, handed down from generation to generation, have been adapted to climates that are about to become extinct." It's a poetic way of saying that the crops we rely on for our daily bread may grow poorly, or not at all, under the changed conditions. "We don't have the option of failing to get our agriculture ready for climate change and ready to feed a growing world population. That means safeguarding [seed] collections," he said. "There is no country in the world, rich or poor, large or small, which is self-sufficient in the genetic diversity of the crops which feed its people. No country can make it on its own."[2]

The Svalbard Seed Vault is a last bastion against the onslaught of crop failure and attendant famine as global climate change unfolds. The world has been debating climate change for some time now, mostly in the context of how expensive gasoline or electricity is going to get and how inconvenient it all is. The impact in the kitchen has barely registered with the average citizen—but it will be big and, quite likely, the greatest impact of all on humankind.

Expected effects on food production under the moderate scenarios for climate change outlined by the Food and Agriculture Organization of the United Nations (FAO) and the Intergovernmental Panel on Climate Change (IPCC) are shown in table 8.

"Climate change will worsen the living conditions of farmers, fishers and forest-dependent people who are already vulnerable and food insecure. Hunger and malnutrition will increase. Rural communities dependent on agriculture in a fragile environment will face an immediate risk of increased crop failure and loss of livestock. Mostly at risk are people living along coasts, in floodplains, mountains, drylands, and the arctic. In general, poor people will be at risk of food insecurity due to loss of assets and lack of adequate insurance coverage," the FAO told the United Nations Climate Change Conference in Bali in 2007. The 2009 World Summit on Food Security in Rome warned bluntly about "severe risks to food security and the agriculture sector" from climate change.[3]

There will be more frequent and more intense extreme weather, with adverse impacts on food production, on food distribution infrastructure, and on livelihoods and opportunities in both rural and urban areas. Changes in mean temperatures and rainfall, increasing weather variability, and rising sea levels will affect the suitability of land for different types of crops and pastures, the health and productivity of forests, the

Table 8 EXPECTED IMPACTS OF CLIMATE CHANGE ON AGRICULTURE

Trend in Extreme Weather and Climate Events	Likelihood	Possible impacts on agriculture, forests, and ecosystems
Over most land areas, fewer cold days and nights, warmer and more frequent hot days and nights	Virtually certain	Increased crop yields in colder environments; decreased yields in warmer environments; increased insect and disease outbreaks. Reduced water supplies from snow melt.
Warm spells/heat waves increase in frequency over most land areas	Very likely	Reduced yields in warmer regions due to heat stress; increased risk of wildfires. Higher water demand, greater scarcity, threats to quality.
Heavy rainfall events become more frequent over most areas	Very likely	Damage to crops; soil erosion, inability to cultivate land due to waterlogging of soil. Risks to safety of drinking water.
Area affected by drought increases	Likely	Land degradation; lower yields/crop damage and failure; increased livestock deaths; increased risk of wildfires; widespread water stress.
Intense tropical cyclone activity increases	Likely	Damage to crops; uprooting of trees; damage to coral reefs and fisheries; disruption to water supplies, power outages.
Increased incidence of extremely high sea level	Likely	Salinization of irrigation water, estuaries, and freshwater systems; decrease in freshwater availability.

SOURCE: Adapted from Food and Agriculture Organization of the United Nations, "Climate Change," March 18, 2009, table 1, www.fao.org/climatechange/49368/en/.

incidence of pests and diseases, the richness of biodiversity, and the health of ecosystems. More arable land will be lost due to increased aridity, groundwater depletion, and rising sea levels, the FAO warns.[4]

"This process of global warming shows no signs of abating and is expected to bring about long-term changes in weather conditions. *These changes will have serious impacts on the four dimensions of food security: food availability, food accessibility, food utilization and food systems stability,*" the FAO states (emphasis added).[5]

Rice is a staple food for half the world's population—and a crop that is particularly vulnerable to global warming. The International Rice Research Institute estimates that for every degree of warming above 32 degrees (89.6° F), there is a 5 percent loss in grain yields caused by temperature-induced sterilization of the spikelets that form the seed. Despite being a tropical crop, rice is extremely susceptible to very hot weather: two or three days of 38–40 degree Celsius heat (100–104° F) at the wrong time can slash the yield of a rice crop to zero.[6]

Katherine Sierra, vice president of sustainable development for the World Bank, has warned that African food productivity could decline by as much as half in the next twenty years or so, while that of Central and South Asia could fall by nearly a third due to warming, changes in rainfall patterns, the impact of pests and diseases, and so on.[7] A study by the International Food Policy Research Institute (IFPRI) concluded that "agriculture and human well-being will be negatively affected by climate change," the main impact being declining yields in the most important crops throughout the developing world, especially in South Asia. Irrigated wheat yields would decrease by 30 percent and rice by 15 percent. This would lead to less food available per capita for humanity in 2050 than in 2000, and twenty-five million more malnourished children. Wheat prices may rise almost 200 percent and rice prices 100 percent. IFPRI estimates that an investment of $U.S.7 billion would be required to offset these losses by helping farmers to adapt to changing climate conditions.[8]

Declines in food productivity of such a magnitude at a time when humanity needs to double production in the face of a dwindling resource base would represent a colossal setback—and are a clear and present danger to planetary security and the human future. In the case of Africa, they could dislodge a flood of millions of refugees in the direction of Europe and the Middle East. In the case of nuclear-armed India and Pakistan, potentially as many as 300–500 million might be driven by the threat of starvation to fight or move, threatening first one another and then Russia, Central Asia, and Southeast Asia.

In China, where the grain bowl region of the North China Plain is already experiencing an alarming drop in the groundwater that supplies two-thirds of the needs of its crops (see chapter 3), rainfall too appears to be failing due to a weakening in the northeast monsoon that blows in from the Bay of Bengal, possibly due to climate change. The phenomenon has awakened memories of grim events: climate analysts suggest that a prolonged age of drought arising from a weak monsoon may have triggered the downfall of China's Tang Dynasty in the ninth century. A collapse in food production in North China today would almost certainly send hundreds of millions of refugees in all directions, but especially north into the Siberian area of Russia, sparking a major escalation in international tensions. "A small Russian population might have substantial difficulty preventing China from asserting control over much of Siberia and the Russian Far East. The probability of conflict between two destabilized nuclear powers would seem high," comments the former CIA director R. James Woolsey in "The Age of Consequences," a think-tank report on the strategic implications of climate change for humanity.[9]

Although some regions may gain longer growing seasons and more rainfall under climate change, there are expected to be no real "winners," the same report asserts: "Any location on Earth is potentially vulnerable to the cascading and reinforcing negative effects of global climate change."[10] Furthermore, the harshest effects may be felt in some of the world's most populous and poorest regions, as table 9 suggests.

The climatic experience of the twentieth century was that rainfall tended to increase over the temperate regions and higher latitudes, and to decrease over the middle latitudes between 10 and 30 degrees. Globally, the area of land classified as dry is thought to have doubled since the 1970s, according to the IPCC. In the decades to come, the proportion of land surface facing extreme drought at any one time is likely to increase, along with a tendency for drying in continental interiors during the summer, especially in the subtropics and low and middle latitudes. Many semi-arid and arid areas (such as the Mediterranean Basin, the western United States, southern Africa, northeastern Brazil, the Australian wheat belt, and the Indian grain bowl) are particularly vulnerable to the impacts of climate change and are projected to suffer a decrease in water resources. At the same time, heavy rainfall events and melting glaciers are very likely to increase flood damage in equatorial regions and high latitudes. "Globally, the negative impacts of future climate change on freshwater systems are expected to outweigh the benefits *(high confidence),*" the IPCC says. "By the 2050s, the area of land subject to increasing water stress due to

Table 9 IMPACTS OF CLIMATE CHANGE ON AGRICULTURE BY REGION

Region	Possible impacts on agriculture, forestry, and ecosystems
Africa	By 2020, between 75 and 250 million people are projected to be exposed to increased water stress due to climate change. In some countries, yields from rain-fed agriculture could fall by up to 50 percent. Agricultural production, including access to food, in many African countries is projected to be severely compromised. This will further adversely affect food security and exacerbate malnutrition.
Asia	By the 2050s, freshwater availability in Central, South, East, and Southeast Asia, particularly in large river basins, is projected to decrease. Coastal areas, especially heavily populated megadelta regions in South, East, and Southeast Asia, will be at greatest risk due to increased flooding.
Latin America	By midcentury, increases in temperature and associated decreases in the water content of soil are projected to lead to gradual replacement of tropical forest by savannah in eastern Amazonia. Semiarid vegetation will tend to be replaced by arid-land vegetation. Productivity of key crops may decrease and livestock productivity decline, with adverse consequences for food security. In temperate zones, soybean yields are projected to increase. Overall, the number of people at risk of hunger is projected to increase.
North America	Decreased snowpack, more winter flooding, and reduced summer water flows are expected to exacerbate competition for overallocated water resources. Moderate climate change is projected to increase yields of rain-fed agriculture 5–20 percent, but with important variability between regions. Major challenges are projected for crops that are near the warm end of their range or that depend on overexploited water resources.
Islands and deltas	Sea-level rise is expected to exacerbate flooding, storm surge, erosion, salinization, and other coastal hazards, threatening the livelihoods of island and other low-lying communities.
Europe	Increased risk of inland flash floods, more frequent coastal flooding, and increased erosion are all predicted, along with glacier retreat, reduced snow cover, and extensive species losses. In Southern Europe, climate change is projected to worsen high temperatures and drought in a region already vulnerable to climate variability, to reduce water availability, and to lower crop yields.

SOURCE: Adapted from Food and Agriculture Organization of the United Nations, "Climate Change," March 18, 2009, table 2, www.fao.org/climatechange/49368/en/.

climate change is projected to be more than double that with decreasing water stress."[11] As evaporation and transpiration increase with warming, irrigators may need up to one-fifth more water just to grow the same amount of food—at a time when cities will be taking water away from farmers, and rivers and aquifers everywhere will be drying up.

SHIFTING GRAIN BOWLS

One of the most profound impacts of climate change will be the migration of the world's main grain-growing regions further north and south of where they are today: out of the United States into Canada, out of China and India and into Russia and Siberia, out of Africa and the Mediterranean and into northern Europe, out of Australia and southern Africa . . . and into the Southern Ocean. The reason for this can be seen from a glance at the atlas: between 15 and 30 degrees from the equator, on both sides, extend broad yellowish bands comprising the world's deserts and drylands—the Sahara, Arabian, Thar, Gobi, and Great Southwestern deserts lie along these latitudes in the northern hemisphere, and the Australian, Atacama, and Kalahari deserts in the southern. These dry regions are produced by Hadley Cells, enormous atmospheric vortexes that corkscrew out from the equator. Initially warm and wet, they shed their moisture as they rise, and then they descend again as dry air around latitude 25 degrees: this falling dry air in which clouds are scarce produces the world's drylands and deserts. Lying in latitudes just beyond these desert belts are the great grain-growing regions, which have the perfect mix of sunshine and moisture to grow the grasses we consume as grain. As the Earth warms and its atmosphere begins to churn more vigorously, however, the Hadley cells will expand, broadening the desert bands and shifting the belts of country suitable for growing cereals farther to the north and south. Middle latitudes closer to the equator will experience more frequent and intense drought, while those farther away will rejoice in longer growing seasons, more rainfall, and higher crop yields.

On the face of it, the gains might be expected to balance the losses—but the reality is that the world will probably lose suitable grain-growing land faster than it can open up new areas in the high north. As we saw in chapter 4, Brazil is one of the countries with large potential for opening up new land to grow food. Under most climate-change scenarios, however, even the vast new territory in the Cerrado is unable to offset the food losses caused by heating and drying in the country's north and east—hence the IPCC's expectation that hunger in Latin America is liable to increase, despite the continent's apparently huge undeveloped agricultural potential.

Nevertheless, the implication is that, under climate change, Canada, Russia, and Siberia will emerge as the world's grain superpowers by the end of the present century. Scientists have calculated that by 2080 Russia/Siberia could gain an extra 40–70 percent of new farming land, while North America may gain 20–50 percent.[12] Such increases would in theory go a long way toward counterbalancing heavy losses in food production in the tropics and subtropics—but unless the food produced in the North is somehow transported and affordably distributed in the deficit regions, the risk of large-scale local famines triggered by climate impacts will be acute. In all of history to date, the existence of vast food surpluses in the North has failed to prevent starvation in the South—and humanity may have to learn to cooperate far better if it is to escape the planetwide consequences of food resource scarcity in its most populous regions.

More important, even success in holding global food production stable at current levels will be no success if the demand for food of the human population overall continues to double.

THE TOWER FALLS

For more than a century the towering peaks of the Himalaya have fired people's imaginations and stirred them to adventure. "Mother of the Snows," Mount Everest stands tall and aloof from the antlike activities of the humans attempting to invade her privacy. But Sagarmatha (Everest) and indeed the entire Himalayan mountain chain now face colossal change at the hands of those same ants—change that will rebound on us. The Himalaya's "eternal" snows enable nearly a quarter of humanity to feed itself, for they constitute a vast reservoir of frozen water, the largest outside the polar ice caps themselves.

This "Himalayan Water Tower" each year blesses the parched plains of the Indian subcontinent, Central Asia, and Western China with abundant flows along ten of the continent's mightiest rivers. "The Tibetan Plateau and the entire Himalayan region in which it stands is perhaps the world's most vulnerable area to climate change," warn mountain and forest researchers Mats Eriksson and Xu Jianchu. "That is because warming increases at higher altitudes and the Himalayan mountain system is the tallest on earth. A scary prospect—for both of the world's most populous nations who lie in and under them and more than a billion people whose livelihoods depend on the water that flows down the mountains." Forty-five percent of the Indus's water and 70 percent of the Ganges' consist of meltwater from this mighty tower. As the climate warms, the river flows will at first increase, causing devastating summer floods, and then, as the glaciers that supply them shrink, the rivers will dwindle. In India the rising flows are already evident, while in the Hindu Kush the rivers are drying up. If they fail, how many of the 1.3 billion who now depend on them will seek to resettle in water-rich Europe, Russia, or North America—and how may such events alter forever the demography and politics of our world?[13]

HOW HOT, HOW DRY?

To gain an inkling of what our children face in their lifetimes, examine the graph based on the work of scientists at Britain's Hadley Centre for Climate Prediction and Research (figure 7).

Though there are rather a lot of wiggly lines, referring to the different climate models and scenarios, the broad message is plain. Taking what has happened to the world's climate in the past and then projecting past and present trends into the future, the British scientists conclude that 40 percent, perhaps even half, of the world may be in the thrall of *regular* drought by 2100. Translating this into food production is not easy, but the Hadley Centre's scientists have attempted it, producing a map that illustrates where it is going to become a great deal harder to grow crops and pastures reliably because of a lack of moisture in the ground. This predicts losses in soil moisture ranging between 20 and 50 percent in northeast Brazil, Central America, West and Southern Africa, the Mediterranean basin, the Middle East, most of the Indian subcontinent, and eastern Australia.[14]

If drought halves local food supplies in key regions, then lack of rain alone may rip away as much as a quarter of the world's food at the very time we will need to double it. These, it should be stressed, are climate models, not crystal balls or prophecies—and the actual outcomes on the farm remain highly uncertain. Nonetheless, they represent a current best estimate of what our children may be facing—and we would be poor parents indeed if we ignored the message that many of them will experience famine and war through our failure to tackle climate change.

These forecasts, it should be noted, are based on a global warming prediction of around 2–3 degrees C (3–4° F) over the present century—an increase that most climate experts assume will occur if carbon levels

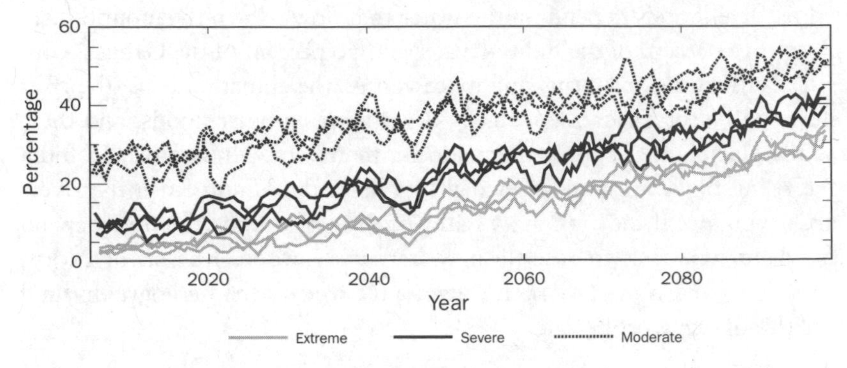

Figure 7 Proportion of the world's land expected to be in drought, 2000–2100. Climate modeling by the United Kingdom's Hadley Centre indicates that 30–50 percent of the Earth's land area may be affected by drought by the end of the twenty-first century. Source: Eleanor J. Burke, Simon J. Brown, and Nikolaos Christidis, "Modeling the Recent Evolution of Global Drought and Projections for the Twenty-first Century with the Hadley Centre Climate Model," *Journal of Hydrometeorology* 7, no. 5 (October 2006): 1113–25.

in the atmosphere are allowed to rise from the current 387 parts per million (ppm) to 450 ppm or more by the midcentury. There are a lot of "ifs" in that assumption, however. The climate may warm faster or more slowly. Droughts may be more frequent and more intense. Because the carbon dioxide we emit today will linger for centuries in the atmosphere, the climate may continue to warm even if there is a slowdown or complete cessation in manmade carbon emissions. Indeed, some researchers argue that global warming is inevitable at any level above 350 ppm—which the world passed back in 1987—and have called for this to be set as the upper limit.

The world currently emits about 50 billion tonnes (55 billion U.S. tons) of carbon dioxide equivalent per year, adding about 1 ppm of greenhouse gases to the atmosphere every four months. At the present rate, it will reach 450 ppm around 2030—a level that can now be avoided only by reducing greenhouse emissions by 80 percent or more immediately. The significance of this number is that, in studying past global climates, a team led by the NASA physicist Jim Hansen has found that the Earth's ice caps began to form when the carbon dioxide level dropped below 425 ppm and its insulating effect diminished. From this they infer that when carbon dioxide levels again pass this number on the way back up, the ice caps will begin to melt entirely (though not all at once). As the policy of most of the world's governments is either to exceed this level or else to do nothing, if the scientists are correct it follows that we are now committed to a course of substantially melting the ice caps—which will eventually raise sea levels, not by a few centimeters, but by up to three meters (ten feet) by the end of this century, drowning large parts of the world's coastal, delta, and low-lying farmland.[15] Hansen and his colleagues conclude that the only safe level for atmospheric carbon is to bring it back down to 350 ppm—about where we were in the middle of the twentieth century, enabling the ice caps to reform. At present levels of atmospheric carbon, Earth is "already in the danger zone." "If the present overshoot of this target CO_2 is not brief, there is a possibility of seeding irreversible catastrophic effects," they warn.[16]

CLIMATE CATASTROPHE

The changes described earlier in this chapter, it should be noted, are based mainly on the rather conservative scenarios for climate change of the IPCC. They do not factor in what scientists refer to as "accelerated climate change"—the events that will unfold if vast areas of tundra melt, releasing their carbon and methane into the air, as deserts expand and likewise release organic carbon from the desiccated soil, as snow and ice caps melt and the Earth soaks up more of the sun's heat, as the oceans reach carbon saturation and can absorb no more, and as vast reserves of frozen methane on the seabed are unlocked, belching explosively into the atmosphere. As each of these fearsome forces is unleashed, the ponderous machinery of climate change gathers momentum—and the temperature in pressure-cooker Earth threatens to rise ever more steeply, five, ten degrees, maybe more.

The planet has witnessed large warming events before—but never during the span of human civilization and almost certainly never over such a brief period. The Carboniferous era, for example, was hot, wet, stormy, carbon dioxide–rich, and abundant with plant life, as we know from the great coal beds laid down at the time—but then there were no people to burn the forests, mine the soil, spread the deserts, or emit more carbon dioxide and so push the process into overdrive. After many millions of years, the Earth's natural dynamics reasserted themselves and cooler phases returned, culminating in the last Ice Ages.

The risks of accelerated climate change are at least being taken seriously, if not by all governments, then at least by those who identify future security threats and plan their defense. The authors of "The Age of Consequences" studied three possible scenarios to make predictions about future security issues: "expected" (an average 1.3 degrees C of global warming by 2040, in line with the IPCC's midrange projections), "severe" (2.6 degrees by 2040, 0.5 meter sea-level rise due to melting tundra and ice caps, etc.), and "catastrophic" (5.6 degrees by 2090, 2 meters or more of sea-level rise). For the severe scenario they envisage, "Crop yields decline significantly in the fertile river deltas because of sea level rise and damage from increased storm surges. Agriculture becomes essentially nonviable in the dry subtropics, where irrigation becomes exceptionally difficult because of dwindling water supplies, and soil salinization is exacerbated by more rapid evaporation of water from irrigated fields. Arid regions in the low latitudes have spread significantly by desertification, taking previously marginally productive crop lands out of production."[17]

Anyone who knows a climate scientist would be aware that climatologists are, in general, far more alarmed about what is happening than major reports such as those by the IPCC have publicly intimated. The prevailing view among climatologists seems to be that we are in for a global warming episode somewhere between "severe" and "catastrophic," which will take several centuries to play out even if we were to cease emitting fossil carbon entirely by, say, 2050. The basis for this concern lies in evidence that carbon dioxide releases are already well above the levels allowed for in the IPCC scenarios, that the ice caps *are* melting much faster than expected, compounding the risk of uncontrollable feedbacks such as the release of methane frozen in tundra and on the seabed (methane has around twenty-three times the planetary warming potential of carbon dioxide).[18] Although individual climate scientists such as NASA's Jim Hansen and Australia's Graeme Pearman have spoken publicly about such risks—at personal cost—many researchers have kept quiet. The IPCC in particular is constrained by the fact that it is an "intergovernmental" body that must proceed on the basis of consensus and with its language moderated by governments, thus passing the science through two kinds of political filter. "Most atmospheric scientists now believe that climate change is going a lot faster and will be far more significant than the public is aware of," Pearman states. One piece of evidence is the dramatic disappearance of Arctic sea ice, which under the "expected" scenario was not expected to vanish until around 2050—but, following the trend of recent seasons, may vanish each summer by as early as 2015. The rapid decay of glaciers in Greenland, the Antarctic, and the Himalayas is another sign. The appearance of large "dead" patches in the oceans growing at 1–4 percent a year, devoid of algal life due to loss of mixing in the oceans as heated water layers stratify, is another disturbing pointer. Pearman contends that even if the risk of catastrophic change is small, the world should know about it in order to take action sooner—rather than when it is already too late. But he admits that saying such things does not make scientists popular.[19]

These extreme climate scenarios have profound implications for our ability to sustain the present food supply—let alone to double it. To the enormity of the challenge of overcoming existing scarcities of water, land, nutrients, technology, and skills they add a layer of difficulty that some experts now view as virtually impossible to solve—at least as regards the maintenance of global food security and societal stability. "The more severe scenarios suggest the prospect of perhaps billions of people over the medium or longer term being forced to relocate. The possibility of such a significant portion of humanity on the move, forced to relocate, poses an enormous challenge even if played out over the course of decades," say the authors of "The Age of Consequences." Such events would manifest, they say, in the form of public rage at governments' inability to deal with the crises, a rise in religiosity and "doomsday" cults, hatred and violence toward migrants and refugees, civil conflicts, and international wars primarily over food, land, and water. In the face of refugee tsunamis and a militant, nuclear-armed Islam, Russia and Europe may become destabilized, the report adds.[20]

Whether or not the ordinary person gives credence to such scenarios, it should be noted that they are becoming staple fare for threat identification purposes in the defense establishments of the United States, Britain, and many other countries whose armed forces are starting to ask themselves, "What is our role in a world where our borders may be besieged by millions of starving, desperate people?" The answer is almost too terrible to contemplate.

RECARBONIZING AGRICULTURE

The only practical solution to climate change is for humanity to cease emitting the gases that insulate the Earth and absorb the sun's heat—carbon dioxide, methane, nitrous oxide, and others—and to make every effort to lock the carbon up again, in soil, vegetation, rock, marine sediment, or deep groundwater. The necessary actions and policies have been well described and are beyond the scope of this book.[21]

Food production is part of the problem and, potentially, an important part of the solution to climate change. Scope exists for humanity to actively adapt food production to a new role mitigating climate change, in addition to keeping us fed.

When the soil is tilled and its microbes disturbed they become more active in breaking down organic matter (carbon) and releasing it into the air as carbon dioxide. When soil is cropped or grazed heavily, it loses its organic matter, and if the process goes too far it becomes desert, releasing its pent-up carbon into the air. When rice is grown in flooded paddies, the bacteria in the mud release methane. When livestock such as cattle, sheep, and goats are fed, especially on poor forage, the microbes in their stomachs emit large quantities of methane as they break down the cellulose in the feed so the animal can digest it. Of humanity's total output of greenhouse gases of 50 billion tonnes (55 billion U.S. tons) of carbon dioxide equivalent per year, the FAO estimates that agriculture produces 5–6 billion tonnes (5.5–6.6 billion U.S. tons) and forestry an additional 8–10 billion tonnes (8.8–11 billion U.S. tons).[22]

Unless we can move quickly to low-carbon farming systems worldwide, doubling the world's food supply implies more than doubling the greenhouse emissions from agriculture alone. Food production and distribution also release vast amounts of carbon into the atmosphere through all the fuels, chemicals, and fertilizers food production uses and the immense distance food is trucked, shipped, and air-freighted to city consumers the world over. These emissions are very hard to reduce or prevent—unless hydrogen or biodiesel fuel is used both for long-distance transportation and on the farm and more food is grown in cities and locally. The main strategies by which agriculture and forestry can help address climate change, as detailed by the FAO, are given in table 10.

The greatest opportunity for agriculture to make a decisive contribution to ameliorating climate change is by "recarbonizing" the soil—in other words, by restoring or enhancing soil's natural fertility and carbon content. In advanced farming systems, especially in higher-rainfall areas, this can be readily achieved by reducing the amount of tillage the farmer undertakes when planting a crop, says Kaddambot Siddique, the head of the University of Western Australia's Institute of Agriculture. "When you disturb the soil less it reduces microbial activity. This in turn means less carbon, water, and nutrients are lost," he says. Minimum-till and zero-till systems, in which farmers sow a new season's crop directly into the stubble of the previous crop or into a sprayed pasture without first plowing, can help to store carbon.[23] Precision agriculture also can make a major contribution to reducing agriculture's energy dependence and improving its climate performance. The same principle applies to

Table 10 WAYS FOR FARMING AND FORESTRY TO HELP FIGHT CLIMATE CHANGE

Sector	Key mitigation technologies and practices currently commercially available	Environmentally effective policies, measures, and instruments	Key constraints or opportunities
Agriculture	Improved crop and grazing land management to increase soil carbon storage; restoration of cultivated peaty soil and degraded lands; improved rice cultivation techniques and livestock and manure management to reduce CH_4 emissions; improved nitrogen fertilizer application techniques to reduce N_2O emissions; growing of dedicated energy crops to replace fossil fuel use; improved energy efficiency; mulch farming, conservation tillage, cover cropping, and recycling of biosolids.	Financial incentives and regulations for improving land management, maintaining soil carbon content, and making efficient use of fertilizers and irrigation	Opportunities: May encourage synergy with sustainable development, reducing vulnerability to climate change, and thereby overcoming barriers to implementation
Forestry	Afforestation; reforestation; forest management; reduced deforestation; harvested wood product management; use of forest products for bioenergy to replace fossil fuel use. By 2030, forest mitigation technologies will include: tree species improvement to increase biomass productivity and carbon sequestration; improved remote sensing technologies for analysis of vegetation and soil carbon sequestration potential and for mapping land-use change	Financial incentives (national and international) to increase forest area, reduce deforestation, and maintain and manage forests; land-use regulation and enforcement	Constraints: Lack of investment capital and land-tenure issues. Opportunities: Help poverty alleviation and provide essential ecosystem services to protect watershed, conserve biodiversity, and advance conservation recreation

SOURCE: Food and Agriculture Organization of the United Nations, "Climate Change Adaptation and Mitigation in the Food and Agriculture Sector," technical background document following FAO consultation in Rome, March 5–8, 2008, table 2, ftp://ftp.fao.org/docrep/fao/meeting/013/ai782e.pdf.

smallholdings, where minimizing soil disturbance, introducing better crop rotations and varieties, and using composted crop wastes can help rebuild soil fertility and lock up more carbon. Another major way of achieving this is through agroforestry—the integration of trees and tree crops with other forms of agriculture and grazing.

Carbon loss from agriculture is greatest in countries where food production is marginal and the land under stress—and one of the most urgent tasks is to spread knowledge among the world's 1.8 billion farmers and pastoralists of how to recarbonize the soil by using minimum tillage, retaining crop stubble, reducing livestock numbers, adopting agroforestry, adding charcoal from burned timber and crop wastes (biochar) to the soil, and other techniques. Terra preta, a technique pioneered by Amazonian Indians centuries ago in which they created a rich, black loam from poor rainforest soil by adding charcoal from cooking fires and food wastes, is one technique now being explored to renovate degraded fields and lock up carbon.[24]

The FAO sees considerable potential for directing money generated by carbon trading in the industrialized world into agriculture, especially in developing countries. "This is a win-win-win opportunity," enthuses Theodor Friedrich, an FAO expert in sustainable production. "We have a chance to slow climate change, help poor farmers make a better living and improve soil health and productivity all at the same time." In another development, a group of farmers calling themselves the Carbon Coalition is campaigning for the use of agriculture as a kind of gigantic sponge to mop up the world's surplus carbon emissions. They argue that the world's farmed soil theoretically has the ability to absorb up to 78 billion tonnes (86 billion U.S. tons) of carbon—a number regarded as overly optimistic by many scientists—and thus can play an important part in helping to wind back climate change. The carbon coalitionist Christine Jones adds that restoring soil carbon has many benefits besides climatic ones: soil that is richer in humus also retains more moisture and nutrients—and therefore yields more bountiful harvests. Richer soil is also less prone to erosion, salinization, and other forms of degradation.[25]

In all, the FAO estimates, food production and forest production, which together produce 13–15 billion tonnes (14–16.5 billion U.S. tons) of greenhouse gas per year, have the potential, with the right practices, to reabsorb 4–18 billion tonnes (4.4–20 billion U.S. tons) per year. At the upper end of this range, it can be seen, it may even be feasible for agriculture and forestry to offset some of the global impact of greenhouse emissions from other human activities.[26] World farming is thus absolutely critical to overcoming the challenge posed by climate change.

This is very hard to achieve, however, and it is here that the impact of the interplay of looming scarcities becomes clear. Just as farmers cannot grow food with water that cities have annexed from agriculture, they cannot lock carbon in good soil that cities and their inhabitants have appropriated for development or recreation. A time may come when it is imperative to reclaim land now diverted out of agriculture to more frivolous uses—and put it to work both for food production and for carbon storage.

Worst degraded are the world's semiarid rangelands, the immense tracts of savannah grassland that support four-fifths of humanity's livestock as well as more than two billion people. These are the lands that are being turned into deserts through overgrazing and unwise attempts at agriculture. Because of their vast area, they have very large potential to soak up carbon from the atmosphere into their soil and vegetation—and this may possibly be achieved by mimicking what nature already does, by allowing grazing animals to roam more freely and limiting their numbers. This insight was developed by the Zimbabwe-born biologist Alan Savory and other pasture scientists. As a young soldier in the African bushveld, Savory observed how wild herds of game never lingered for long in one place, but always moved in search of sweeter grass. Human-tended herds, in contrast, often stay in one place until it is completely eaten out. Seeking an explanation, Savory found that when plants in dry areas are grazed too hard, their root systems contract, which reduces the amount of carbon in the soil and makes it prone to blow or wash away. By moving humanity's herds constantly, like the wild game herds, and allowing the grasses and shrubs to recover, we can rebuild the damaged soil of the world's rangelands, Savory argues, yielding more food and wealth sustainably, protecting wildlife, and at the same time fighting climate change.[27]

The Australian farming-systems expert David Kemp says that returning grasslands to native perennial species, instead of introduced grasses, would greatly improve their yields, reduce erosion, save water, and lock up far more carbon. In practical terms, if Australia and China alone were to restore their degraded rangelands by reducing livestock numbers as little as 10–20 percent and allowing native vegetation to regrow, the restored rangeland would lock up an extra 400 million tonnes (440 million U.S. tons) of carbon per year, while increasing total production by around 20 percent from animals that were no longer starving and pastures that were no longer degraded.[28]

Though it is an unappealing thought, every time you draw breath you inhale a small quantity of cattle belch and fart—so vast are the emissions from the world's 3.6 billion ruminant livestock. There are several ways to reduce the amount of methane emitted by livestock, such as developing better-quality pastures and feeds that are more easily digested by animals, with less waste of gas and energy. A second option is to manage the actual microbes in the rumen (first stomach) of the animals, selecting for those that are most efficient at turning fodder into feed for the animals—or even to develop improved microbes using genetic techniques. A third option is to shift the diets of meat-producing animals away from grain and back to high-quality grazing, as discussed later. Fourth, we can shift global meat production away from ruminants, which release so much methane, and toward pigs, poultry, and fish. And fifth, we can eat less meat.

EATING LITE

A vital strategy in fighting climate change is for the whole of humanity to eat at a much lower dietary level, for the sake of the climate as well as our personal health. Besides restraining greenhouse emissions this will help to reduce obesity, degenerative disease, malnutrition, and premature death in all societies—so that, besides helping to save the planet, consumers will also undoubtedly be saving their own lives, good health, and taxes.

A study by the British researcher Tara Garnett, "Cooking Up a Storm," estimates that feeding the United Kingdom's 61 million consumers produces 43.3 million tonnes (48 million U.S. tons) of greenhouse gases, and this accounts for 19 percent of the country's emissions.[29] Dutch research shows that meat, dairy, and soft drinks account for two-thirds of the greenhouse content of the typical Western diet, and it is hard to avoid concluding that if we are to take personal responsibility for our individual impact on the climate, here is a good place to begin.[30]

Humans have been habitual meat eaters for two million years or more, and strict vegetarianism is unlikely to catch on universally or to be enforceable—although it is becoming somewhat more popular in affluent societies. Some authors are already contemplating meat rationing and other more punitive measures. Tara Garnett thinks that consumers are unlikely to respond to education or awareness campaigns alone—and suggests that they may need both regulation and price signals to induce them to change their ways.[31] As the impact of the coming famine begins to bite universally and the necessity becomes clearer, however, there is the potential for an act of intelligent free choice by people willing to reduce their intake of high-energy foods in order to help stabilize both the climate and global food security. These may sound a little like wartime measures—but the alternative to not adopting them may well be wars.

WHAT CAN I DO ABOUT IT?

1. Rebalance your diet toward foods that have a smaller "carbon footprint," such as fruits and vegetables.
2. Consume meat, oils, sugar, soft drinks, and dairy products more sparingly.
3. Favor foods that are seasonal and are grown locally and with low-energy inputs.
4. Favor foods produced by systems known to enrich the soil. Seek consumer information on how your food was grown.
5. Support government policies that promote "carbon farming" and the recycling of organic wastes.
6. Waste no food personally, but recycle and compost anything left over for return to the soil or to agriculture.

TEN

ELEPHANTS IN THE KITCHEN

> Sustainable human development will occur when all humans can have fulfilling lives without degrading the planet.
>
> **—Global Footprint Network**

At some point in the twenty-first century, the century of peaks, the world is going to experience "peak people."

If the rate of growth of the human population in the late twentieth century were to continue indefinitely, it has been calculated, within a few centuries the whole of Earth's land surface would be packed solid with human beings.[1] Those who rationally consider the issue, setting aside their personal preferences and cultural baggage, quickly realize that such a scenario is impossible, that even maintaining such rates of growth for a few more decades is highly undesirable. At some point, the population must start to decrease—especially if humans are to enjoy reasonable living standards.

"Any dispassionate analysis of the increases now occurring in the world's population must inevitably result in consternation and a degree of alarm," the agricultural scientist Derek Tribe once wrote. "If allowed to continue unchecked, exponential growth—albeit at a declining rate—will ultimately spell disaster for the planet Earth. Finite resources such as water, soil, plants, forests, animals, energy and minerals, upon which all human life depends, will inexorably be destroyed, degraded, exterminated or used up."[2]

The planet cannot sustain an infinitely growing population of people, especially on Western diets. The real issue is how "peak people" will occur, when, and whether it will be before some other critical resources such as land, water, nutrients, or energy begin to give out—or as a painful consequence of these scarcities. Both natural and human histories are replete with chilling reminders about what happens to almost any creature or society that outruns its resources: they fight, they run, they sicken, they starve. Jared Diamond in his compelling book *Collapse* argued that the civilizations of the Maya, the Anasazi, the Pitcairn and Easter Islanders, and the Greenland Norse fell because they ran through their environmental resources and then failed to adapt to the changed conditions. Among Diamond's primary drivers of past civilization collapses were loss of forest, soil degradation, poor water management, overhunting, overfishing, introduced species, increased population, and increased food demand—all phenomena that characterize the modern world. To these we have added human-induced climate change, a toxic environment, and emerging shortages of both energy and photosynthetic capacity. As a consequence, Diamond argues, "our world society is presently on a nonsustainable course."[3]

A contrasting perspective was advanced by the Danish political scientist Bjorn Lomborg, who put the case in 2001 that the big environmental problems—pollution, water shortages, deforestation, species loss, overpopulation, malnutrition, and diseases such as AIDS—were due primarily to poverty and chiefly confined to the poorer regions of the world. These problems could be overcome by economic and social development, he reasoned. Lomborg went on to argue that global problems such as peak oil and global warming had been overstated and that governments had overreacted to them with inappropriate policies. Furthermore, he pointed out, life expectancy, food intake, and global prosperity had all continued to rise.[4] Lomborg's opinions, it should be noted, were based on trends at the end of the twentieth century, before high food prices had flagged the scarcities that were building up in the background of world agriculture. Nonetheless, his opinions remind us of the necessity of caution toward the more extreme scenarios, as well as emphasizing that an essential component of the solution lies in economic progress, cooperation, education, and social enlightenment.

Since Lomborg published his views in 2001, the declining availability of land, water, nutrients, fish, technology, and energy and the impact of climate changes have become far more evident, and it is no longer easy to dismiss them as exaggerated. It is clear that if we are to double food output by midcentury, we will be doing so in a context in which the fundamental ingredients for such an increase are likely to become critically scarce.

Table 11 GLOBAL POPULATION PROJECTIONS, 1950–2050

Year	Population
1950	2,535,093
1960	3,031,931
1970	3,698,676
1980	4,451,470
1990	5,294,879
2000	6,124,123
2005	6,514,751
2010	6,906,558
2015	7,295,135
2020	7,667,090
2025	8,010,509
2030	8,317,707
2035	8,587,050
2040	8,823,546
2045	9,025,982
2050	9,191,287

SOURCE: Population Division of the Department of Economic and Social Affairs of the United Nations Secretariat, *World Population Prospects: The 2008 Revision*, http://esa.un.org/unpp/index.asp?panel=1 (as of June 3, 2009). UN figures are often revised; these were checked on June 3, 2009.

Recent projections for the world population provided by the United Nations indicate that "peak people" is unlikely to happen—barring a catastrophe—in the first half of the century. By 2050, based on various assumptions, the UN's population division projects human numbers to be slightly less than 9.2 billion and still rising (see table 11).

The first thing to notice about this table is the marked deceleration in the rate of population growth that is expected to occur as living standards rise, life expectancies increase, and birth rates fall. This is already happening: after gaining 840 million people in the 1980s and 830 million in the 1990s, global population was expected to swell by 780 million in the first decade of the twenty-first century. On the face of it, it appears that the wave-peak in population *growth* may already have passed sometime in the 1990s[5]—though not the peak in actual numbers, as we are also living longer.

According to the UN's projections, this slowdown is expected to continue, with population growth falling to 760 million more people in the 2010s, 630 million in the 2020s, 500 million in the 2030s, and 360 million in the 2040s. This is due mostly to a steady and ongoing drop in the birth rate: in 1950 there were 37 births for every 1,000 people, by 2000 there were 21, and, if the trends hold up, there will be just 13.6 by 2050.

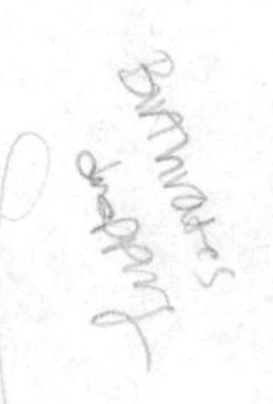

Put another way, the average woman of 1970 had 4.32 babies in her lifetime. The woman of 2005 had 2.56, and by 2060 the average woman is expected to have only 1.85. In the 2060s or thereabouts, if nothing else changes, we will pass the crisis of peak population and human numbers will embark on a gentle decline. So the second thing to notice is that, regardless of culture, race, or religious affiliation, the women of the world are reducing their fertility, in most cases on their own initiative and with little or no reference to men. On the surface the main reasons for this are thought to be increasing affluence, better education and employment opportunities, improved child health care (resulting in lower infant death rates), and urbanization, although enforced approaches such as China's one-child policy have also played a role.[6] Apart from the many direct reasons for people to have fewer babies, it is tantalizing to speculate that something unspoken may also be happening at the species level that is prompting more and more women to pull back instinctively from the dangers of overpopulation and to ignore the cries of those politicians, journalists, and business and religious leaders—nearly all males—still urging them to accelerate the headlong rush to overpopulate.

In coming decades the greatest population growth will be in the world's less developed regions, which by 2050 will be home to 85 percent (7.9 billion) of humanity's 9.2 billion people. By then, however, fertility rates will be much closer: 2.02 children per woman in developing regions and 1.79 children per woman in developed regions. Life expectancy for people in the developing world may average ten years longer than it is today, and five years longer in the developed world, however:[7] this will prolong the time during which the Earth has to carry a population of humans larger than its systems can support without, in some cases, collapsing.

Not all scenarios for future population are as rosy as the UN projections, however. For example, Gwynne Dyer argues in *Climate Wars,*

> It is extremely unlikely there will ever be nine billion human beings on the planet. It's not just that there is no obvious way to feed the next two-and-a-half billion. In the near future global heating is going to start depriving us of a large and steadily increasing proportion of the food supply that supports the present 6.7 billion. There will be famines and a great many people will die. So while we work frantically to get our greenhouse emissions to

> zero, we also have to find ways to avoid the wars that would increase the deaths by an order of magnitude—wars that would also cripple our attempt to avert the runaway climate change that would cause megadeaths later on.[8]

In Dyer's view, there is a dangerous cycle leading from famine to war to planetary overheating and, depending on how far this goes, possible extinction should the Earth flip back into one of its previous toxic states, like the Great Death of the Permian (discussed in chapter 6).

On the food front I am more optimistic than Dyer is, given all the opportunities we have at least to maintain current output in the face of scarcity by reducing waste, recycling resources, developing alternative methods such as biocultures and algae farming, and reinvigorating the research enterprise for both high-intensity and smallholder agriculture. We need to acknowledge, however, that the mainspring of the human population explosion has been our success at raising food output—and that the opposite also holds true: if we fail to consistently repeat the miracle, it will soon bring our numbers down again, one way or another.

This second "inconvenient" truth makes it all the more important for humanity to discuss the issue of population reduction openly, frankly, freely, honestly, and without fear of the reaction of those who are unable, for one reason or another, to grasp the enormity and speed of what is happening. In view of the possible consequences of failure to secure the food supply—up to and including nuclear wars, mass starvation, and floods of refugees numbering in the hundreds of millions—to indulge continued social, media, and political censorship on this issue is to court disaster. All people should be free to talk about the matter, to form their own conclusions, and to act as they think best. Defenders of unrestricted reproduction have hitherto claimed the moral high ground, asserting that nothing is more important than the bearing and raising of children. This is a view that no longer stands firm in a century where its consequences may be starvation, flight, and slaughter on an almost unimaginable scale: people who advocate more births need to explain why they also implicitly advocate more killing, hunger, and misery. Our reluctance as a society to face up to the issue of population is the first of the two elephants in the kitchen—huge, ominous, but insufficiently acknowledged presences in our immediate future.

There are no uniform prescriptions to solve the population problem. Different societies will adopt different solutions—a one-child policy may suit communist China but would not sit well with most North Americans or Europeans, who prefer that such decisions be made by informed consent on the part of thoughtful individuals. Besides, argues the historian Matthew Connelly, attempts to enforce population control are usually cruel and often backfire.[9]

How the decision to have fewer children is made is far less important than the fact that it *should* be made. Nonetheless, some leaders—for instance, in Japan—are still urging their aging citizenry to have more children without regard to the consequences. Left to pursue its current natural course, the Japanese population would shrink from 127 million to 89 million by 2055, providing exceptional leadership by example in a world in which the resources to support high living standards for a burgeoning population are becoming scarce. But old ideas die hard, and many governments still attempt to bribe their citizens to have more babies. France, Russia, and Scandinavia, for example, all encourage their citizens with free child care and other subsidies. In one Russian province, people are given Wednesdays off as the Day of Conception—while citizens who conceive on Russia Day can compete for cash prizes and four-wheel drives. Spain and other countries deliver a cash bonus for every child born, while South Korea underwrites artificial conceptions.[10] Almost every government in the world provides economic subsidies and policy incentives to its citizens to increase population—both because of a conviction that there are votes in it and because of the deeply embedded economic doctrine that growth demands that we produce more and more people in order to consume the goods made by the society and generate the taxes necessary to support high government spending, and also because we need more young citizens to support the aging ones. Governments, in other words, are driven by primitive notions of what is required for economic growth, as well as by (poorly informed) public and business expectations, to subsidize population growth. This flies in the face of recent economic experience that many countries—examples being Japan, South Korea, and Singapore—have achieved high levels of income, tax revenue, and living standards by investing in knowledge, technology, and education rather than by the crude expedient of merely trying to multiply their people without regard to the long-term consequences.

The danger in this policy of subsidizing reproduction lies in the fact that most of the world's governments are, unwittingly, contributing to one of the major drivers of the coming famine. It is time to reconsider all such beliefs and policies before we reach the point at which human demand for food collides with resource scarcities.

Hearteningly, however, there is a generational element at work that is stronger than these outdated views: elderly citizens come from a generation that mostly believed that it was sound policy to try to swamp competitors—national, religious, or ethnic—with sheer numbers. This may have had a rational basis a century or more ago, but the world has since changed, and many are unaware of how much or how rapidly this has happened and cannot revise their opinions to encompass it. It is not they, however, who will have to live in the uniquely perilous world of the coming famine. Middle-aged and younger people, and especially young women, are far less committed to the idea of having many children than their parents and grandparents were; many defer the experience and an increasing number wish for none at all. As mentioned earlier, young women the world over, regardless of nationality, culture, or religion, are voluntarily and perhaps instinctively reducing their fertility. The potential for a worldwide consensus led by these young women about having fewer babies and raising them better remains one of our best hopes for the future—a voluntary, rational, and willing reduction in human numbers rather than an involuntary collapse.

Besides freely discussing and understanding the need to limit birth rates and relying on young women to lead us to a more sustainable human population, there is a third thing humanity can do in the short run to reduce its fertility voluntarily: end poverty. The United Nations's Millennium Development Goals urged us to do this out of a sense of disgrace that wealthy humanity still tolerates such unfairness, but the rich countries broadly ignored the call to conscience. Now there is a more compelling reason even than shame: self-interest—our very survival and avoidance of awful consequences. One of the striking features of modern demography is that the better off people become, the fewer babies they have. It begins with the biological fact that as living standards rise, fewer children die in infancy. The wealthiest countries tend to have the lowest birth rates, especially where the wealth is evenly distributed and immigration rates are low.[11] Then it progresses to the instinctually understood point that, from the standpoint of perpetuating your lineage, it is better to raise and educate one or two children well than five or ten children poorly,[12] and this seems to be where many young women are heading nowadays. As a rule, birth rates are highest in the least-developed and poorest societies, fall fastest in the newly industrializing and urbanizing world, and are consistently lowest in the most developed and affluent societies. It follows from this that prosperity is the most potent contraceptive in the human medicine cabinet. Give people enough of it and they will autonomously decide to have fewer babies without having to be argued, cajoled, or forced into it. Besides its altruism, enlightened self-interest would suggest that we need to end poverty out of a wish to protect humanity from the inevitable consequences of unrestrained population growth.

A positive and direct way to limit world population is to share knowledge—especially of sustainable food production and infant nutrition and care—more freely with poor societies so they can use it to work their own way out of poverty, as societies usually do. They will then naturally choose to lower their own fertility. Such a straightforward and practical approach will have the threefold effect of improving world food security, reducing hardship, and restraining population growth.

OF FEET AND FOOTPRINTS

The coming famine is less about sheer human numbers than it is about the size of our feet—or at least the size of the "footprints" we all leave on the world's natural environment as we seek to satisfy our needs for food and other resources. Our ecological footprint is the amount of resources it takes to maintain each of us on the planet. As we have seen in earlier chapters, this includes quite mind-boggling quantities of water, land, nutrients, oil, and other forms of energy—some of which are not easily renewable. The concept was developed by William Rees and Mathis Wackernagel in the early 1990s as a way of estimating how many people the Earth can carry at various standards of living.[13]

Their Global Footprint Network (GFN) estimated that to support the *average* citizen of Planet Earth required 1.8 hectares (4.4 acres) of land and sea in 2003. Based on living standards, the average American needed 9.4 hectares (22.2 acres), the typical Australian 7.8 hectares (19.3 acres), the average Swiss 5.0 hectares (12.4 acres), and the typical Chinese 2.1 hectares (5.2 acres).[14] Using the same calculus, the GFN has found that, if present rates of growth in our demand for resources continue, by 2050 we will need two entire planets' worth of resources to satisfy that demand.

But we don't have to follow this path, Wackernagel adds. At present, he cautions, it takes the Earth one year and four months to regenerate what humanity consumes in twelve months—a phenomenon known as "ecological overshoot," which means that we are living beyond our means (see figure 8). As this gap between production and consumption widens, the risks of famine and conflict grow. "Humanity is living off its ecological credit card," Wackernagel says.[15] The analogy is all the more vivid when we have all seen how living beyond our means on toxic debt destroyed the world's financial markets. What was lost in the crash of 2008 was

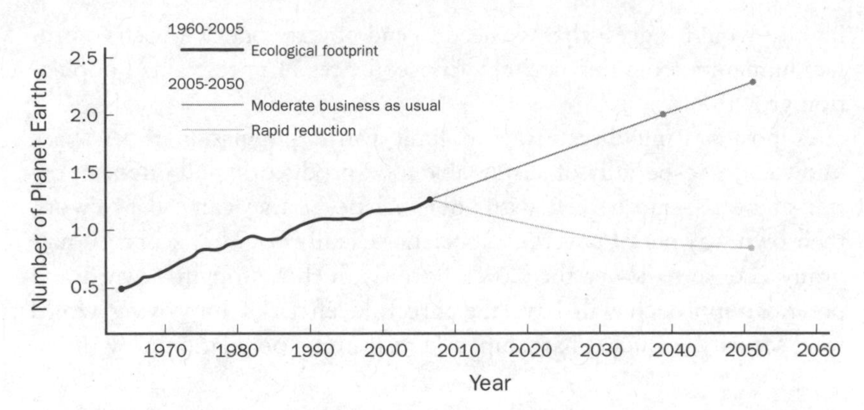

Figure 8 Ecological overshoot, the extent to which our demands exceed the renewable resources of Planet Earth. The darker line indicates that continued moderate growth in our demands would require two planets' worth of resources by 2040. Rapid reduction in demand would make us sustainable by 2050. Source: Global Footprint Network, "World Footprint: Do We Fit on the Planet?" 2009, www.footprintnetwork.org/en/index.php/GFN/page/world_footprint/.

mainly money, however. What is being lost now is far more serious: the Earth's ability to support our survival with sufficient food.

The GFN's work demonstrates that if everyone consumed at the levels of the average citizen of Bangladesh or Sierra Leone, we would still be well within the Earth's capacity to support us in the long term. By contrast, if all humans adopted a U.S. living standard, we would need four planets. Continuing increases in our global demand for resources will inevitably impose major decisions on the coming generation about what we eat, how much we eat, and how we produce the food—as food production is one of the most resource-depleting things we do.

In a recent essay, the paleontologist and author Tim Flannery wrote:

> There is no real debate about how serious our predicament is: all plausible projections indicate that over the next forty to ninety years humanity will exceed—in all probability by around 100 per cent—the capacity of Earth to supply our needs, thereby greatly exacerbating the risk of widespread starvation, or of being overwhelmed by our own pollution. The most credible estimates indicate that we are already exceeding Earth's capacity to support our species (termed its biocapacity) by around 25 per cent. With global food security at an all-time low, and greenhouse gases so choking our atmosphere as to threaten a global climate catastrophe, the signs of what may come are all around us.[16]

UNSATISFIED APPETITES

One of the very first things people do when their incomes start to rise above poverty levels is to improve their diet, which usually means eating more meat, oils, fish, eggs, and dairy products. After a time this hunger for protein eases off, but usually only after people start to eat far more than they need, and so become fat and sick. Illustrating this, meat consumption in the United States (123 kilograms or 271 pounds per person) and Europe (74 kilograms or 163 pounds per person) has risen slowly or not at all over the past generation, whereas in China it has tripled to 54 kilograms (119 pounds) per person and India has only just begun the ascent from a very low level indeed.[17] This growth is all attributable to families having enough income to afford high-energy foods, and it happens in every country as it climbs the living standards curve. How quickly it occurs depends to a considerable degree on how slowly or rapidly the global economy and the economies of poor countries expand. The global recession slowed things down temporarily—but, inevitably, the trend resumed.

Spurring the growth in consumption is the media. Whereas a century ago most people would have been content with their local and largely vegetable diets, nowadays affluent lifestyles are blazoned in the media for all to see, and the glamorizing of high-energy diets—advertised in fast-food outlets, on billboards, and on TV—urges people in every city on Earth to overconsume.[18] Another frequently overlooked driver of the rush to high-energy, wasteful diets that breed ill-health are the social welfare policies of rich countries, which today tend to encourage reliance among the urban poor on junk food and prepackaged meals, rather than on producing and preparing their own healthy sustenance.

The result of these pressures has been a global feeding frenzy that has, in less than a generation, completely overwhelmed centuries of thrift, careful husbanding, home-growing, and recycling of food resources. Most people now feel entitled to high levels of protein and fat consumption and, if they cannot obtain them in their birthplace, are willing to move across nations and continents. The increased global mobility of the ravenous consumer—migrant, refugee, business traveler, expatriate, backpacker, or vacationer—has added an intractable new dimension to the developing crisis. Wherever they go, these affluent twenty-first-century nomads tread more heavily on the Earth in their demand for high-energy foods, unconsciously imperiling the tenuous resources of the local poor and, equally unconsciously, spreading the toxic message that unrestrained, promiscuous consumption is OK.

The second elephant in the kitchen is thus the demand for food, and in particular consumption of livestock products—a second huge presence in our lives we don't much care to talk about. Various Food and Agriculture Organization (FAO) estimates suggest that global meat output could climb by as much as 180 million tonnes to reach 465 million tonnes (512 million U.S. tons) by 2050.[19] As we have seen, this will require us to find an extra two thousand cubic kilometers of water from somewhere to grow the feed for the animals, an extra two billion tonnes of grain. In short, we need to discover two more North Americas to produce all that fodder; however, anyone who has visited Google Earth can probably attest that there are few undiscovered continents left. With its flair for understatement, the FAO told the World Food Conference in 2007, "Satisfying the increasing demand for animal food products, while at the same time sustaining the natural resource base and coping with climate change and vulnerability, is one of the major challenges facing world agriculture today."[20]

It is hard to envision what all this means for the world's food supply, but some estimates from the U.S. Department of Agriculture give us an impression. Every American "consumes" an average of 753 kilograms (342 pounds) of corn a year, which is mainly used to nourish the livestock that produce meat, eggs, and dairy products. Imagine how it will affect world grain supplies were Chinese consumers (who outnumber Americans four to one) to increase their feedgrain intake *sevenfold* to achieve a similarly meat-rich diet—and what may occur should equally numerous Indian consumers abandon centuries of vegetarianism, adopt Western foods, and increase their corn consumption massively. In case this should be thought improbable on cultural grounds, it is worth bearing in mind that, in the 1930s, 97 percent of the diet of the average Chinese was composed of grains and vegetables; today it is just 67 percent.[21] Given a chance, most people eat meat.

In India, a recent study predicted, "Over the next two decades, India's middle class will grow more than tenfold to become the world's fifth largest consumer market with almost 600 million consumers."[22] By 2030 the Indian population will overtake that of China, and—economic circumstances permitting—the country will have a greater number of affluent consumers than America and Europe combined. Such growth is expected to place very significant strains on Indian agricultural capacity and self-sufficiency, quite apart from the probable impacts of climate change. Even though there are significant parts of the Indian population who do not eat beef or pork, many nevertheless consume sheep and goat meat, milk, poultry, and fish as well as oils and sugar.

FACING THE ELEPHANTS

These two huge demand-side "elephants" of population growth and hunger for protein are the greatest drivers of the coming famine. Both elephants are poorly acknowledged: even raising the issues provokes such a vehement reaction from minorities, conservative religious groups, and industry lobbyists that politicians are, for the most part, frightened to do so. As a result, public policy in many countries is being bullied and censored into a state in which it will deliver the exact opposite of what is necessary for the survival of humanity through the midcentury food crises.

As we have seen, there are many things that can and must be done on the supply side of the global food system to overcome the looming resource scarcities. But the impact of climate change in particular means that they are unlikely to be sufficient to keep all humanity fed through the peak in numbers and consumption. If we are to avoid global instability, action is also essential on the demand side—by finding ways to encourage the falling trend in birth rates and to rein in our appetite for high-energy protein. This is where solving the crisis comes down to the individual. We can and should no longer expect governments and science alone to "fix it."

The University of Manitoba resource ecologist Vaclav Smil has calculated that two-thirds of all the water used by the United States to produce food goes into meat, egg, and dairy production. "Consequently, if Americans were to reduce their consumption of beef, pork and poultry by a third they would still eat nearly 60 kilograms of meat a year per capita, but such a dietary modification would save annually 120 to 140 cubic kilometres of virtual water," he says. Even with such a reduction, the average American would still consume 100 grams (0.22 pounds) of meat a day—more than that needed for a healthy diet. "Rational food production and healthier eating in America and Europe could thus eliminate the need for at least 250 cubic kilometres of virtual water every year. This would be more than twice as much as is saved annually by the international trade in food and feed, and it would suffice to produce 200 to 250 million tonnes of cereals that could be exported to water-deficient nations, nearly doubling the mass of grain that is now traded annually worldwide," Smil says.[23]

Whether affluent consumers will happily reduce their meat intake is a moot question. A recent investigation of intensive animal production in the United States by the Pew Commission hints, however, that this is possible: it found that the industry was a threat to the health of the community generally; polluted air, land, and water; and was cruel to animals and harmful to rural society. The commission made the point that most city consumers are completely unaware of the conditions under which their meat is produced—clearly implying that once they find out, they may be a good deal more reluctant to eat it. It recommended that all intensive confinement of animals be phased out within ten years, adding, "There is increasing urgency to chart a new course. Our energy, water, and climate resources are undergoing dramatic changes that, in the judgment of the Commissioners, will require agriculture to transition to much more biologically diverse systems, organized into biological synergies that exchange energy, improve soil quality, and conserve water and other resources."[24] The report did not recommend an end to meat, dairy, or egg production or consumption—but rather a shift to a less intensive, more balanced and sustainable system.

This points the way to the future of meat consumption—more sparing, less energy- and water-intensive, and kinder to animals, people, and the environment. In particular it opens the way for the renewal of the livestock grazing industries of the world's rangelands on a more sustainable footing, providing meat and other products that are clean, of higher quality, use less energy, and have less impact on the environment. Forging the social consensus necessary to bring this about in the wealthy nations, however, let alone in the emerging economies whose citizens are sampling the delights of protein food in abundance for the first time, has yet to begin.

Even if North Americans and Europeans halved their meat and dairy consumption, the saving could be completely swamped by the demand from six hundred million newly affluent Indian and Chinese consumers. To this the answer is that the Western diet, high in fats, chemicals, sugars, salt, carbohydrates, and overprocessed foods, is increasingly recognized as the starting point for most of the ill-health and deaths in Western societies. If India, China, and other emerging economies wish to avoid this, along with the astronomical health-care costs that accompany it, they will encourage their citizens to consume protein far more sparingly by educating them about the risks. Societies still close to their rural roots are more keenly aware of what makes a healthy diet than affluent societies, so persuading them may prove less difficult than it appears.

As we approach the midcentury peak in food demand it becomes ever more vital to talk openly about such matters, to let people the world over draw their own conclusions, and to find ways to moderate our consumption in line with what the Earth, and our own bodies, can support.

WHAT CAN I DO ABOUT IT?

1. Feel free to discuss issues of population and how it can be restrained. Do not allow censorship, prejudice, or intimidation to prevent this.
2. Make a good decision about your own fertility based on your understanding of the global challenges all humans, including your children, now face.
3. Think about your diet and how it can be made less destructive to the environment.
4. Teach your children to respect food and never to waste it.

A PLEA FOR HONESTY

When people call for better family planning or help for women seeking to control their own fertility, they are sometimes misrepresented by critics as urging "population control." It is therefore necessary to state that this book does *not* advocate population control: it advocates individuals making rational and well-informed decisions about their own fertility based on accurate information about what is really happening in the world.

When people call for lower consumption of meat and dairy, they are sometimes misrepresented by industry interests and the media as advocating vegetarianism. Likewise, this book does *not* advocate imposed societywide vegetarianism. It advocates increased consumption of vegetables, fruits, and grains and decreased consumption of high-energy foods in a balanced, healthy diet of the sort your great-grandmother would probably have approved of.

On both issues, informed free choice is more likely to obtain results than enforcement. Humans have been overcoming threats to our survival through consensus, common sense, and cooperative action for more than a million years, and there is no reason for us to stop now. But in order to take collective as well as individual action, we must first clearly understand and discuss the issues.

ELEVEN

A FAIR DEAL FOR FARMERS

> The food crisis of today is a wake-up call for tomorrow.
>
> **—UN Secretary General Ban Ki-moon**

"We are alarmed that the number of people suffering from hunger and poverty now exceeds 1 billion. This is an unacceptable blight on the lives, livelihoods and dignity of one-sixth of the world's population. The effects of long-standing underinvestment in food security, agriculture, and rural development have recently been further exacerbated by food, financial, and economic crises, among other factors. We must collectively accelerate steps to reverse this trend and to set the world on a path to achieving the progressive realization of the right to adequate food in the context of national food security." Thus the World Summit on Food Security summed up in November 2009 the situation facing the world's poorest people, and the continuing failure of governments to remedy the problem. However, like so many previous well-intentioned declarations, the countries of the world set themselves neither firm deadlines nor specific monetary targets nor measurable goals, leaving a largely rhetorical commitment.[1] World farm commodity prices, which had rocketed up almost 70 percent during 2007 and the first half of 2008, came tumbling down again as recession gripped the world economy in late 2008. Their fall was nowhere near as steep, however, as the slump in world share values, real estate, or metals prices, and in most countries retail food prices fell little, if at all. Then, as the world pulled out of the economic slump, food prices began to climb again.

"Global food prices have eased significantly from their record highs in the first part of 2008," the analyst Alex Evans noted in his report on the global food outlook for Britain's Royal Institute for International Affairs. "However, this does not mean that policy-makers around the world can start to breathe a sigh of relief. Looking to the medium and longer term . . . food prices are poised to rise again." The reason, he said, was that in predicting a return to the thirty-year trend of falling food prices, the Food and Agriculture Organization of the United Nations (FAO) and others had "largely overlooked" the emerging shortages of water, land, oil, fish, and technology as well as rising demand.[2]

These all point to the conclusion that—for the time being at least—the era of cheap, abundant food is over. In the three decades between 1974 and 2005, real food prices fell by around 75 percent, says the International Food Policy Research Institute (IFPRI). Real food prices then almost doubled before easing again with the economic crash.[3] As with oil, increased volatility in food prices is a symptom of approaching scarcity. Although food prices are also bound to fluctuate widely in response to short-term signals such as the size of harvests, the state of the global economy, or the price of oil, a growing consensus among global analysts is that for many years to come, food prices are likely on the whole to rise, reflecting greater uncertainty than in the past generation.

The claim is commonly voiced that the world has plenty of food, more than enough for everybody; it is just very poorly distributed. There is truth in this but also great oversimplification. The surplus food grown in rich countries has never fed more than a tiny proportion of the world's poor, and if this problem is to be solved the poor must be given the knowledge and means to feed themselves. It is not the total supply of food that needs to be increased so much as poor people's access to it. Yet the food produced in rich countries is too expensive, the countries that grow it show little inclination to pay for large-scale delivery to hungry regions (and indeed, prefer to burn it in vehicles or throw it in the garbage), and it is currently highly reliant on limited resources such as oil, phosphorus, freshwater, and land. Furthermore, wealthy countries continue to pursue trade and subsidy policies that obstruct trade and distort prices, thus exacerbating food insecurity, hunger, the risk of refugee crises, and conflict.

BLOCKING FOOD

"We must correct the present system that generates world food insecurity on account of international market distortions resulting from agricultural subsidies, customs tariffs and technical barriers to trade, but also from skewed distribution of resources of official development assistance and of national budgets of developing countries," declared Jacques Diouf, the director general of the FAO, when he announced the 2009 World Food Summit. Progress toward the liberation of world agricultural trade is essential, argues Ajay Vashee, the president of the International Federation of Agricultural Producers—a world body claiming to represent 600 million of the world's 1.8 billion farmers. IFPRI has also called for the elimination of all restrictions on agricultural trade, the enactment of fair trade rules, and greater regional openness. The German aid organization Welthungerhilfe argues bluntly, "Fair trade is a must for developing countries; the European Union and the industrialized countries must cancel their import restrictions and abolish agricultural export subsidies."[4]

If poor farmers in developing countries can sell surplus produce to wealthier countries, the argument runs, they can obtain better incomes and so lift themselves out of poverty, in the process improving both domestic and global food security. Furthermore, if developed countries abandon their domestic farm subsidies, the world prices of farm commodities will improve and with them the incomes of poor producers. Round after round of world trade talks, however, have yielded meager progress or else collapsed because wealthy nations preferred to protect their own—often inefficient—farmers by subsidizing domestic food production and restricting imports, caring little whether this policy inflicted poverty and death elsewhere.

"Subsidized agriculture in the *developed* world is one of the greatest obstacles to economic growth in the *developing* world," the National Center for Policy Analysis, a U.S. think tank, explains. "In 2002, industrialized countries in the Organization for Economic Cooperation and Development (OECD) spent a total of $300 billion on crop price supports, production payments and other farm programs. These subsidies encourage overproduction. Markets are flooded with surplus crops that are sold below the cost of production, depressing world prices. Countries with unsubsidized goods are essentially shut out of world markets, devastating their local economies. Moreover, farm subsidies lead to environmental harm in rich and poor nations alike." Poor farmers, it says, lose $24 billion per year in income due to rich country subsidies. The International Monetary Fund has argued that the elimination of farm subsidies would boost the world economy by $128 billion a year.[5]

Yet the situation continues to deteriorate: in 2008, many countries took steps to shield themselves against the global food price spiral, which had the effect of worsening global food security. By restricting or banning grain exports, imposing price controls, providing consumer subsidies, stockpiling grain, and rechanneling food through social protection programs, they exacerbated the global shortage.[6] Like farm subsidies, this illustrates how comparatively minor and well-intentioned domestic measures can combine to have a reverberative impact on the world's ability to feed itself—an impact that is seldom taken into consideration when the local measure is introduced. The peculiar irony of this situation is that the agricultural and trade policies of developed countries are significant drivers of the global food crisis—and the policies they then adopt to overcome its rebounding impact on themselves serve only to make matters worse.

Freeing up world trade in food and abolishing subsidies, economists argue, would allow the production of staple foods to move to wherever in the world it can be most efficiently done, where the climate is most favorable and the land and water most available. This would enhance global food security. Sugar production, for example, would tend to move from Europe to Brazil and other developing countries, which can do it more cheaply and better. This would benefit not only poor countries but also consumers everywhere. It would also send direct price signals to farmers worldwide—undistorted by political intervention—telling them what to grow, how much, and where best to grow it. On a planet where our footprint has quite possibly already exceeded the available resources, it is becoming increasingly urgent to use the Earth's total food production resources—the "trophosphere"[7]—far more efficiently, sustainably, and sparingly than we do at present.

It must also be clearly understood, however, that freeing world farm markets will tend to focus production in the hands of the most efficient farmers (large or small)—to the detriment and ultimate failure of the less efficient, whether they live in rich countries or poor ones and whether they practice high-energy or low-input farming. Many of the advocates of advanced farming methods skip over or completely ignore their social consequences. Freeing markets and introducing more productive technologies usually involves large-scale movements of people out of rural areas and into cities. The choice facing national governments is thus between, on the one hand, periodic global food crises and greater instability and, on the other hand, having to assist the adjustment of many farmers out of agriculture and into other worthwhile livelihoods, with the clear-sighted intent to improve the overall performance of the worldwide food system and its ability to feed humanity as a whole from an increasingly limited resource base.

The pain caused by this displacement of rural people can be alleviated or overcome in various ways:

- by employing displaced rural people in new food industries, such as urban horticulture or biofarming, where their skills can be used productively,
- by employing them at places all along the food value chain,
- by employing them in work to improve water and nutrient recycling, in conservation, in infrastructural development, in carbon sequestration, and in large-scale agroforestry,
- by coupling rural adjustment policies with policies for the development of related secondary and tertiary industry,
- by paying farmers directly for their role as stewards of the nation's and world's water, soil, and wildlife, thus providing them with a partial living, and
- by developing a fairer global system of payment to farmers (see the discussion later in this chapter).

In the short run, liberating international food trade may seem like an ideal doomed to fall forever foul of domestic politics. Countries perennially make the error of confusing food self-sufficiency with food security. Politicians constantly appease noisy rural protectionism, ignoring the greater benefits to be had from more efficient food production and use of resources. Consumers and taxpayers are constantly asked to prop up inefficiency, often for purely sentimental reasons. But if the world is to avoid the coming famine—and rich countries to escape being swamped by hundreds of millions of immigrants and refugees from regions where food production has failed—freeing world farm trade may well emerge as the preferable and pragmatic option.

Free farm trade may also be an essential strategy for combating climate change. People now understand that human land use can prevent wild animals and forests from migrating to more favorable regions as the climate changes; the same is true of agriculture, where trade barriers, city sprawl, and other distortions prevent food production from moving naturally to the most climatically favorable regions or to the most efficient farming systems. The FAO comments that open economic trade, reduced subsidies, and income diversification, coupled with potential new income for rural populations from providing environmental services (such as carbon sequestration, reforestation, species protection, and sustainable energy production), are among the important steps for successful adaptation and mitigation of climate change.[8]

There is no better time to achieve trade reform than in an economic downturn, when the benefits of enhanced trade flow swiftly through the economies of the participating countries to generate stronger economic activity and restore employment. Yet this is also a time when it is hardest for politicians to win the domestic argument for free trade, as they are under siege from a citizenry fearful of further job losses or company failures. Fearing this, the G20 leaders stated in their November 2008 pledge to tackle the world economic crisis, "We underscore the critical importance of rejecting protectionism and not turning inward in times of financial uncertainty. . . . [W]e will refrain from raising new barriers to investment or to trade in goods and services, imposing new export restrictions, or implementing inconsistent measures to stimulate exports. . . . [W]e shall strive to reach agreement that leads to a successful conclusion to the [world trade negotiations]. We instruct our Trade Ministers to achieve this objective."[9]

UNBLOCKING FOOD

The key to world trade reform lies in each country having a clear grasp of the damage it does to itself and its citizens through trade barriers and subsidies, and the benefits to be gained from removing them, argue the economists Andy Stoeckel and Hayden Fisher. The true costs and benefits of different trade policies are usually obscured in the clamor of vested interests and domestic politics, leaving society uncertain about the best policy to pursue.

The key is transparency, with the public receiving an unbiased, accurate, and trustworthy account of the costs and benefits of trade policies to the whole economy. Stoeckel and Fisher say that this can be achieved in four ways:

1. Give the public all relevant information on the policies in place and how they work.
2. Establish a visibly independent and publicly trustworthy body to review the policies.

3. Make this body's findings public, so government is under pressure to take on board the findings of these reviews and is unable to cave in to vested interests without being seen to do so.
4. Evaluate all policies in an economywide way (i.e., not just in terms of their benefits to a particular group or region), so they are consistent with one another and work together for the greater good.[10]

Transparency works in several ways: it identifies the national interest for all to see and informs and educates both government and the public about it. It exposes vested interests that would profit at the expense of the many and weakens their influence over the political process. It helps sometimes-competing sectors of the economy to find common cause in improving trade and other policies and develops coalitions for reform. And it helps to deliver greater economic certainty, less distortion, and a sounder investment climate.

Stoeckel points to the Productivity Commission of Australia as a body that achieves policy transparency by being independent of government, producing highly credible reports, and taking an economywide approach. Governments find it difficult to ignore such rational, balanced, and well-publicized advice. "The key to world trade reform lies in each country coming to its own, clear understanding of the benefits it can gain from opening its markets," Stoeckel says. "You can't make a proper decision in the national interest if you can't measure the national interest—and this is what an independent commission can do by evaluating the competing arguments fairly and objectively. People whose countries have high trade barriers have a right to know what they are losing by not having low ones."[11]

GLOBAL FOOD: THREAT OR OPPORTUNITY?

In barely a generation food has become globalized, meaning that not only is it traded internationally but diets and culinary habits once restricted to particular countries or ethnic groups have become popular elsewhere, and in some cases everywhere. What began as the export of cooking styles such as American fast food, Chinese food, and French or Italian cuisine has flourished into an enormous trade in raw and processed foodstuffs as giant food manufacturers and supermarket chains seek to source out-of-season, exotic, convenient, and cheap produce from around the world.

"The private sector is now an important player in the changing world food situation, and the related innovation and research system," comments Joachim von Braun, the director general of IFPRI.

> The corporate food system is increasing its market share in developing countries, where urbanization is creating new retail food consumers. Indeed, the fastest-growing element in the food chain in developing countries is the supermarket sector with growth rates in sales exceeding 20 percent per annum in some countries. In addition, food manufacturing and processing are on the rise, as urban consumers demand more processed foods. Industry is increasingly looking at the millions of small farmers and poor consumers as customers. In some regions, such as East Asia, small farmers are getting a foothold in this changing food chain, but the smaller they are, the higher their unit transaction costs to participate in the food chain. Companies along the food chain are becoming locally increasingly concentrated . . . , highly coordinated along the food chain . . . , and more global in their operations. . . .
>
> [T]he world's farmers, of which 85 percent have less than 2 hectares, must link in new and efficient ways to input, processing, and retail industries to capture a fair share in the value chain. The concentration rate is a matter of concern to many. However, at a global scale, even the allegedly highly concentrated retail sector's top five companies . . . do not yet capture more than 14 percent of the market. . . .
>
> At the same time, rising consumer incomes and changing lifestyles are creating bigger markets for high-value agricultural products like fruits, vegetables, fish, and meat. The growing markets for these products present an opportunity for developing-country farmers to diversify their production out of staple grains and raise their incomes. Annual growth rates on the order of 8 to 10 percent in high-value agricultural products . . . are a promising development . . . , as the production, processing, and marketing of these products create a lot of needed employment in rural areas.[12]

The globalization of food, driven by the giant food processors and supermarket chains, is one of the main reasons why consumers in developed countries and large cities have enjoyed such variety of cheap food for so long. Using their market power, big food firms are able to extract the most competitive prices for produce, which has the effect of driving down returns to farmers around the world. Big firms do this by seeking out the best and most efficient suppliers, whether in the local country or anywhere in the world, and then binding them to a contract that specifies exactly what produce is to be delivered. The sheer market power of these large buyers puts pressure on all other farmers to either match the price or go out of business. Such a system works to the advantage of the supermarket and its customers in the short term, giving them low-priced and generally good-quality food (though the quality issue is contentious)—but often to the disadvantage of farmers, who are weak sellers trapped between lower prices for their produce dictated by large, muscular food firms and the rising costs of fuel, fertilizer, and other inputs dictated by equally muscular industrial corporations. In extreme cases, entire local farm industries close down. Even when food production moves offshore, the benefits generally go only to a few highly efficient producers, and generally only until their costs begin to rise—and the supermarket chain discovers a cheaper source of supply in another less-developed country. Other farmers in the developing country may benefit indirectly from the local economic growth caused by exports, but may also be disadvantaged or even put out of business by the competitive pressure from their successful peers.

This system produces profits for the supermarkets and food processors and encourages economic efficiency in agriculture but is widely disliked by farmers around the world, who carry most of the risk and are usually forced to accept lower commodity prices and incomes. Large-scale farmers who are in a position to profit most can afford to adopt the best new technologies and more sustainable systems—but this does not apply evenly across the whole of agriculture, forcing many producers to operate unsustainably, mining their soil, overstocking their grazing lands, or failing to control pests. Consumers have mixed feelings, being appreciative of low-cost, high-quality produce year-round but concerned over issues such as country of origin, food hygiene and safety, environmental sustainability, industrialization, and chemicalization of the food chain.

From the standpoint of world food security, the globalization of food in its present form poses a number of risks that are insufficiently recognized.

- Sourcing produce from the cheapest supplier internationally can undermine local food security.
- Because of the large number of "food miles" involved in acquiring produce globally, globalization of food is vulnerable to increases in the cost of freight and fuel scarcity and is a major contributor to global warming, which in turn damages agriculture. Excessive transportation thus undermines food security.
- Because it specifies the quality and visual appearance of produce so tightly, globalization of food leads to great waste of farm produce that does not meet its standards.

- The need to overstock supermarket shelves in order to entice customers also involves wastage of food that has passed the use-by date and promotes a culture of waste.
- Because of falling returns and rising input costs, farmers in general have less capital to invest in adopting more sustainable practices and new technologies. Thus, globalization of food encourages greater economic efficiency among a few suppliers, but often at the expense of the environment, natural resources, rural society, and overall food security.
- Globalization of food favors large suppliers over small ones. This has social consequences in rural areas, causing the loss of much farming talent, youth, and enterprise.
- Globalization of food favors subsidized farmers over unsubsidized ones, which distorts the economics of the food supply and encourages bad behavior by governments.
- The price pressures globalization of food exerts can force farmers into unsustainable use of land and water and overuse of fertilizers and pesticides, inflicting damage on the agro-ecosystem and exposing consumers to a potentially toxic cocktail of chemicals.
- Although they extract large profits, food manufacturers and supermarkets rarely invest in agricultural research or sustainable agricultural systems to help farmers become more efficient or sustainable. They therefore reap the profits without regard to the ultimate impact on the Earth's resources or assisting farmers to protect these.
- The large chemical, biotech, and fertilizer companies carry out excellent research aimed at improved farming methods and yields, but this has a bias toward their own balance sheets rather than the health of the agro-ecosystem, public good, consumer wishes, or rural society.

The British environmental writer George Monbiot is a prominent critic of the modern agrifood chain. "Big business is killing small farming," he says.

> By extending intellectual property rights over every aspect of production; by developing plants which either won't breed true or which don't reproduce at all, it ensures that only those with access to capital can cultivate. As it captures both the wholesale and retail markets, it seeks to reduce its transaction costs by engaging only with major sellers. If you think that

> supermarkets are giving farmers in the UK a hard time, you should see what they are doing to growers in the poor world. As developing countries sweep away street markets and hawkers' stalls and replace them with superstores and glossy malls, the most productive farmers lose their customers and are forced to sell up. The rich nations support this process by demanding access for their companies. Their agricultural subsidies still help their own, large farmers to compete unfairly with the small producers of the poor world.[13]

Monbiot concludes that fair trade is possibly the only way that small producers, who are efficient in their use of scarce resources, can survive.

Many agribusiness firms, such as Monsanto, Bayer, and Syngenta, do indeed claim to be addressing some of these issues and to be helping poor smallholder farmers in the developing world to raise yields.[14] They will undoubtedly make a major contribution to yield improvement and the reduced use of water, energy, land, nutrients, and pesticides, as well as enhanced climate adaptability, in many countries over the coming decades. They will naturally focus on the regions and the crops that are most profitable to themselves, however, rather than those that most enhance food security for humanity. To head off the coming famine, the world will need both their contribution and that of a reinvigorated public research effort. "The pattern of low global investment in agricultural research and development has contributed to slower growth in agricultural productivity. Unless the world addresses these challenges, the livelihoods and food security of millions of poor people, as well as the economic, ecological, and political situation in many developing countries, will remain at risk," warns the IFPRI's Joachim von Braun.[15]

The more enlightened supermarket chains and manufacturers are responding to some of these challenges—especially to their consumers' demands for more organic and ecologically sustainable produce. Supermarket chains, like farmers, are not immune from the necessity of having to match the prices of their competition if they want to stay in business. Nor are they immune to the food preferences of their customers.

Yet the buck for looking after the world's food-producing system does not entirely stop with the supermarket, agribusiness firm, or food manufacturer. It stops with the person who eats the food: with you and me. By paying cheap prices for food, as we all like to do, we undermine the capacity of farmers to take care of the resource base and make farming more sustainable. We promote the mining of soil, water, nutrients, fossil energy, and fish—not to mention harming human lives and rural communities and, often, being cruel to animals. By being uninformed and unaware of a food-production system from which most of us are now almost completely divorced, we unwittingly encourage global prices for food that do not reflect its true cost to the environment or to the people who grow it. This is what most needs to change.

It is time to recognize that the globalizing of the food-purchasing and food-delivery system is having negative as well as positive impacts on the trophosphere, the world's total agricultural resources, and thus on food security. This is potentially a serious market failure, as it prices food at less than the cost to the planet of producing it and encourages the long-term degradation of the very resources needed to sustain our nourishment into the future. This situation cannot last and demands reform as urgently as do world trade and the financial markets. Various methods have been suggested for addressing this situation, all of which have benefits and drawbacks.

1. Price: through the payment of fairer prices to farmers by consumers, supermarkets, and food processors, which would enable them to protect the food resource base and invest in new technologies, at the same time limiting demand for the most resource-intensive foods.
2. Subsidy: by government payments to farmers for their stewardship of soil, water, and biodiversity, separate from their commercial food production.
3. Regulation: by limiting by law those practices or technologies that degrade the food resource base and/or rewarding those that improve it.
4. Taxation: by imposing on all food a resource tax that reflects its true cost to the environment to produce, and by reinvesting the proceeds in more sustainable farming systems, rural adjustment, and enhanced resource management.
5. Education: by educating the public about how to eat more sustainably, and by educating industry about sustainability standards.
6. A combination of several of these measures.

Whatever the preferred solution—and different countries will favor different approaches—there probably needs to be a combination of sanctions and rewards, market mechanisms and laws, public awareness and industry codes of practice to achieve a system that avoids the further degradation and depletion of the Earth's food-producing resources.

To meet humanity's food needs to 2050, the FAO estimates the world needs to invest an additional US$83 billion a year (on top of the current global investment of US$142 billion). In the same report, however, the FAO recognizes that investment will not be forthcoming from farmers or the private sector unless it is profitable for them. In other words, the present global food marketing and pricing system, including consumer attitudes, is a major barrier to world food security.[16]

CREATING WISER CONSUMERS

A vital key to solving the problems created by the globalization of food lies in evolving a global system that allows consumers and agribusiness to reward farmers who invest in more sustainable food systems—large and small—and more productive technologies, yet does not distort commodity prices, foster destructive practices, or inhibit trade. Some countries have sought to address this by paying farmers directly for their stewardship of land, water, and wildlife, while insisting that they still earn a commercial living by selling their crops and livestock on the open market. If the food chain has become globalized, it may be time to consider a global scheme along these lines that protects the resources needed to produce food and provide environmental services to all of us.

One way or another, humanity will have to face the true cost of growing its food. This is unavoidable.

The Australian marine scientist David Bellwood observes that the issue of our age is that most humans are now entirely separated from their natural environment. He sees the destruction taking place in the oceans as due primarily to the fact that people are oblivious to the results of their actions—and that the damage is "out of sight, out of mind": only fishers and researchers get to see and understand it firsthand.[17] Consumers seldom connect a small piece of fish on their plate with the collapse of a fishery—yet we are the ones who initiate the price signals to those who ruin it. The same is true of the devastation and waste taking place all along the terrestrial food chain, which this book has sought to illustrate: the well-off consumer gobbles up 2,000 tonnes (2,200 U.S. tons) of water and 66 barrels of oil per year in the form of food, along with hundreds of tons of soil and tons of wasted nutrients; we make deserts by subjecting farmers and herders to relentless economic pressures; we drive farmers to overfertilize and overchemicalize by demanding large, flawless produce; we squander vast quantities of the food they do produce; and we help pollute the Earth's natural systems with nutrients and poisons. We do this, indirectly, every time we shop for food. And most of us don't have a clue, or give a damn, about it.

This suggests that one of the most effective countermeasures we can take against the coming famine lies in the realm of education. By helping consumers, especially children, to understand how food is produced we can enable them to make wiser and less damaging choices, to waste less food, and to conserve nutrients, soil, and water. Since food, and its attendant issues, is so pivotal a matter to our personal and collective survival, it is fair to propose that an entire year of primary schooling, around age six or seven, should be dedicated to how food is grown, what makes it grow, and what is good to eat and what isn't. Children with their naturally open minds are fascinated by sprouting seeds, by farm animals and crops, by basic cookery, by nature. Teaching food also offers an enjoyable, hands-on way to teach other subjects, such as mathematics, the environment, science, history, society, and geography, which can sometimes be a little dry and unengaging when divorced from the personal experience of the primary schooler. Food is a way to spice and enliven the instruction of many other subjects and show them from a perspective the student can identify with.

There are urgent reasons to do this. As many as three out of every five of the children presently coming through school in the well-off world are destined to die as a result of what they eat, from the host of chronic disorders that result from overnutrition.[18] Many of the diseases of affluence—heart disease, cancer, and diabetes—are also now spreading like wildfire in the developing world as diets change there, too. We accept the principle of giving children vaccines to prevent illness and untimely death. Perhaps we should also vaccinate them with greater knowledge about food, so they can live longer, healthier, and less costly lives. Perhaps we should vaccinate them against the mining and destruction of the Earth's food-producing systems and the wars that follow by teaching them more about how to grow and consume food sustainably. Perhaps we should vaccinate them against waste and pollution of our natural ecosystems by teaching them how this can be avoided and about the virtues of recycling.

Introducing a "food year" in the world's primary schools would be among the most powerful measures for raising a generation of educated consumers, a healthier generation, a generation of more productive and sustainable farmers, and one of better-informed politicians, bankers, TV chefs, magazine editors, and business people with a sounder grasp of the consequences of their actions. In a world divided by so many things, it would create a common experience for people in all countries and societies, based around an issue fundamental to our existence and a universally shared experience: food.

Teaching food is acceptable to every country and creed, adaptable to every geographic, climatic, or cultural circumstance. You would obviously teach food differently to Inuit children than to young Pitjantjatjara in the Australian desert, Hmong in Vietnam, or kids in New York, Shanghai, Dusseldorf, or Jogjakarta—but the principles would be the same: never take more from the natural world than it can replenish. Cherish your water, your energy, your soil, your plants and animals, your nutrients. Understand their needs. Have respect for them, for they are what keep you alive. Eat sparingly, and only what is good for you and for the Earth.

Although the curriculum may vary from culture to culture and school to school, the basic science of how to grow food sustainably, the science of protecting the trophosphere and the science of health and diet are the same for all of us. Working this science into a format suitable for use in schools and then disseminating it to the world's education systems for local adaptation and sharing via schools, nongovernmental organizations, the Internet, and mass media to reach a global audience is not an insuperable task. Most of the information already exists and is readily available. The main challenge is one of universal dissemination. To sum up, if we are prepared to globalize food, then we must also be willing to globalize knowledge about it so we can produce and consume it more wisely.

For developing societies, a focus on teaching how to produce food efficiently and reduce waste would provide a tremendous boost to the education of young women, farmers, and livestock producers, especially in places where primary education is just starting to penetrate. It would help to lower child mortality and reduce poverty. The accumulated knowledge of agricultural and dietary science and the Green Revolution can be distributed far more widely and productively than it is at present. For high-consuming societies, a single year of education about food may be insufficient, so a second year could be added in junior high or high school, around age thirteen. (This is just when most kids, including future presidents and leaders of industry, switch off science because they find it too dull. Food could help retain their interest—and might result in saner policies when they grow up and take power.) Food should also be taught as a component of university and technical courses in science, environmental management, medicine, nursing, architecture, city planning, and the like, far more than it is at present.

The value of a direct appeal to the world's young cannot be overestimated. So fast does human knowledge grow nowadays that it is often the young who educate their parents rather than the other way round. If we are to depend on the world's young women to change global attitudes toward fertility, then relying on our children to change our attitudes to food, waste, and unsustainability may also be a good strategy. Furthermore, it will send the supermarkets and agrifood chain direct signals to which they can respond. If they are sincere in their wish to serve consumers, then these corporations can spare a part of their profits to help fund this worldwide dissemination of sustainable knowledge about food. It is a job well tailored to such global enterprises and can be regarded as one of their ethical obligations to the society in which they prosper.

GLOBAL RECESSION: A FOOD-LED RECOVERY?

The global economic slump proved a mixed blessing for world food security. On the one hand, it brought farm commodity prices down somewhat. This cut farmers' incomes but did not everywhere translate into cheaper food prices. On the other hand, it did little or nothing to change the major factors that constrain future food production, the drivers of the coming famine explored in this book. According to the World Bank, the global economic crisis added an additional hundred million people to the ranks of the world's very poor and, if prolonged, will add many more still.[19]

The longer-term consequences of the global economic recession harbor some downsides for food security. These include increased food price variability as farmers cut back on inputs such as fuel and fertilizer and, in particular, increased hunger, poverty, and hardship for the poorest citizens on the planet.

An increase in poverty is liable to be accompanied by an increase in the birth rate in poorer countries and poorer sections of society—and there are already signs this may be happening.[20] This could, for a time, slow or even reverse the global trend toward a lessening in population growth. Food demand overall may actually increase: in recessions consumers tend to buy fewer luxuries, and food increases as a proportion of their household spending. While the strongly growing demand for food and meat in the newly affluent classes in India and China will probably ease slightly, it is unlikely to reverse unless these giant economies, too, cease to grow. As world economic growth resumes, the strongly rising demand for protein from newly industrialized nations will resume.

In the case of water, recession is likely to mean that governments the world over will continue to underprice water, encouraging waste rather than risking the wrath of water users. Hard times in rural areas may accelerate the movement of poor people into the cities, leading to greater pressure by those cities on rural water supplies used to grow food. Recession is also liable to delay the introduction of recycling and sensitive urban water design as well as the costly technologies needed to increase farm water use efficiency. The net result will be greater global water insecurity, with obvious consequences in coming decades for food production.

In the case of farmland, the sudden drop in farm prices means several things. First, the sag in farmers' terms of trade means that the adoption of sustainable farming systems will be delayed, leading to continued or increased soil and water degradation. Second, poorer farm profitability will tend to discourage or retard the opening up of new land for agriculture or the rehabilitation of degraded farmland. Whether lower land values will lead to more farmland being snapped up by developers is unclear. On balance, recession is liable to exacerbate the dearth of good farmland.

In the case of nutrients, the price of fertilizers is driven chiefly by the price of oil and gas and, in fact, tends to rise faster. After rising sharply in previous years, world fertilizer prices eased somewhat in 2008 and 2009. Because phosphorus and potash are both based on finite mineral sources, however, and nitrogen fertilizers depend largely on finite natural gas, lower prices are likely to mean that global resources will be consumed more quickly, leading to the prospect of further nutrient price volatility as fears about potential shortages grow again. This increases the urgency of recycling nutrients now wasted on- and off-farm.

In the case of oil, cheaper prices—though a relief for farmers and transportation operators—mean the same: we will burn through the world's known oil reserves at a faster rate and, since exploration for new oil depends on the profitability of oil companies, the gap between supply and demand is likely to widen. Several commentators have observed that this raises the specter of a second, even more dramatic, oil price shock as economies begin to work their way out of recession.[21] If this happens it will in turn drag food prices sharply up again. At the same time, recession will probably slow investment in biofuels, as these will be less competitive with oil, releasing more land and farming muscle for food production.

For the world's fish stocks the impact of recession is mixed: since fish is already relatively expensive compared to other forms of protein, consumers may well limit their demand for it, and this will tend to slow catch rates, buying time for governments and fishers to introduce more sustainable wild-harvest practices. Equally, the recession may slow the rate of investment in new aquaculture enterprises and the production of farmed fish. This will increase the pressure on land-based protein production in the decades ahead.

For climate change, the recession augurs a bleaker outlook. In the medium term, governments the world over will be extremely reluctant—despite their best intentions—to burden their struggling economies with high carbon prices or taxes. Investment in clean technologies may be deferred. The temporary cheapness of oil and coal may cause usage to increase despite reduced industrial demand, raising global greenhouse emissions and accelerating warming and ocean acidification.

A hazard of the recession is that, as their revenues dwindle and budgets fall into deficit, governments will cut back on agricultural research and overseas aid, reducing or eliminating potential solutions to the coming famine. Agricultural research lacks popular support among voters. It is easy and politically cheap to cut when the financial going is hard. The logical way to avoid this, as the world rides out the downturn, is to rechannel some of its defense expenditure into war prevention via enhancing global food security.

The bottom line is that the global economic recession may slow but will not fundamentally alter the factors driving supply and demand for food in the decades ahead. Indeed, it may make some of them worse. As the British analyst Alex Evans observes, "There is therefore a real risk of a 'food crunch' at some point in the future, which would fall particularly hard on import dependent countries and on poor people everywhere."[22]

Agriculture, as we have seen, is the foundation on which a valley, a nation, or a region builds its prosperity and stability. Today we have globalized food and are approaching a global agricultural system. The best investment the world could possibly make in tough times lies in securing the world food supply and in using this to regenerate stalled economies and trade through this most vital of human activities. Investment in food security costs, relatively, so little that we often treat it as unimportant. If we hope to avoid the coming famine, however, making this investment is also the one thing we absolutely cannot fail to do.

Only through the application of knowledge can we be sure there will be enough food in the future. So let us eat and prosper together.

TWELVE

FOOD IN THE FUTURE

> The year is 2085 and Yasmin's teacher has taken the small class to the local museum. Their assignment is to discover how their ravaged world has come to be. As they enter the darkened central chamber, a single ray illuminates an object displayed on a blackened pedestal, a thing so forbidding that its vague form—let alone the messages it contains—sends a chill of horror through the awed children. As they draw closer they begin to feel its power. Closer and they can at last make out what it is, this fount of all the ruin, the suffering, the hunger, the loss.
>
> It's a cookbook.

Those raised on conventional museums are accustomed to the guns, swords, religious symbols, and legal documents with which our thuggish ancestors imposed their views on their fellows, or died in the attempt. The victors had the privilege of framing history the way they liked. In the food wars to come, however, there will be no victors—only victims, and they will see the matter differently. The typical early twenty-first-century cookbook, with its gorgeous illustrations, elegiac combinations of the failing fruits of the Earth with those that cost us the climate, water, soil, and our safety to produce is, unambiguously, a recipe for disaster on a planetary scale.

For decades cookbooks have encouraged us to cherish the fantasy that the abundance is endless, cheap, and without penalty: very rarely have they advised us of the true ecological costs of eating. Often they imply that our small personal indulgences do not matter in the greater scheme of things. That as long as our table groans with delicacies, we can overlook the consequences for others, our children and their descendents especially. By the middle of this century, the traditional cookbook could be as much a symbol of devastation and misery as the most dangerous of military weapons or the gaudiest of national banners. Like Nero's lyrics to a burning Rome, the traditional cookbook is a hymnal to an age of indulgence that is costing us the Earth.

It should be reasonably plain by this stage that the cookbook, besides containing mouth-watering recipes, is also—unintentionally—one of the devices that is fueling the coming famine. There are many others, of course, but the cookbook serves to illustrate the point. More than anything, it evokes the ancient human instinct to fill the belly while the good times last, for famine is written in our genes as the thing *most* to be feared.

Though few seem aware of it, we humans—no matter where we were born—are all originally the children of drought, famine, and drylands. Our species evolved in an era when the Earth was hot, the African woodlands parched and dying, the Mediterranean bone dry and glittering with salt crystals, while volcanoes crackled along the heaving floor of the Great Rift Valley. Our closest relative, the chimpanzee, took her leave of us, clinging to the forests. We stood up and, alert to peril, moved out into the open grasslands. Learned to cooperate. Learned to kill and avoid being killed. Learned to cook. Ate meat, not fruit. Constructed the Pyramids, the Taj Mahal, Tupperware, penicillin, pantyhose, and the Joint Strike Fighter. Wrote sonnets and sitcoms. Composed concertos and cookbooks.

The lesson of the last four million years is that we are nothing if not survivors. But, like our upright ancestors gazing watchfully across that parched savannah, we like to clearly see trouble coming and to understand its nature in order to avoid it. Our success as a species is founded on our ability to foresee danger—and respond to it before it is too late.

The coming famine is a planetary emergency. It is so because by the midcentury no country, no region, and no person will be unaffected. Indeed, its impact will be felt by most life forms on the planet, one way or another. Even those who do not personally go hungry will feel the food price and other economic impacts, the higher taxes, the loss of landscapes and species, the contamination, the tidal surges of immigrants and refugees, and the iron lash of conflict.

The coming famine is not, it must be stressed, a single event like the Irish potato famine. It will probably be a nonlinear crescendo of events brought on by growing regional scarcities of land, water, nutrients, fuels, technology, fish, and skills—scarcities that are already interacting with and amplifying one another. These resonate with rising human numbers, an increasingly erratic climate, and our seemingly ungovernable appetites. It is the confluence of colossal needs with titanic shortages.

The coming famine is also a planetary emergency because it will take more than the combined will and goodwill of the governments and potentates of the world to solve. It will demand the active and willing participation of each individual, every community, all farmers and consumers. It is a challenge at the species level like no other, even climate change. For the first time in history we must all pull together if we are all to pull through. From this hunger and its wider consequences there are *no* hiding places. As Ajay Vashee, the president of the world farmers body the International Federation of Agricultural Producers, observes, "Global challenges need global solutions."[1]

Global solutions will not be found, however, until most people come to a sensible appreciation of the jeopardy in which we stand—and the challenge of explaining this to seven or eight billion people is large enough on its own. Persuading them all that it is in our common interest to act together, instead of as warring tribes and competing nationalities, defies all prior human politics. It is the absolute test of our self-lauding title *sapiens*.

In each previous chapter dealing with a scarcity, various possible solutions have been reviewed. Many more, due to space limitations and their technical nature, have been omitted but are nonetheless important. It is the principles that count, and among these there are five measures that, above all, we may take to avert the coming famine. They are each practical, affordable, and achievable. Individually, they are just the sort of challenges we thrive on.

A HEALTHY WORLD DIET

The first challenge is to develop a diet for everyone that is sparing of energy, water, land, and other inputs and has minimal impact on the wider environment. In the past half century the world has undergone a culinary revolution, and now the challenge is to embark on a second. A sensible—and delightful—way to do this involves increasing the proportion and diversity of vegetables in the diet.

Vegetable production yields more food for a given area of land than legumes, cereals, or meat production, says the World Vegetable Center.

Table 12 PRODUCTIVITY OF FOOD RELATIVE TO LAND AND WATER INPUTS (GLOBAL AVERAGES)

	Yield (ton/ha)	Water productivity (kg/m3 or g/l)
Vegetables		
Cabbages/brassicas	22.5	11.3
Carrots and turnips	22.3	9.8
Cauliflowers and broccoli	18.7	
Chilies and peppers	15.3	
Cucumbers and gherkins	17.3	
Eggplants	15.7	
Okra	6.5	
Spinach	15.7	3.4
String beans	8.7	
Tomatoes	27.3	5.9
Legumes		
Lentils	1.0	
Soybeans	2.3	
Chickpeas	0.8	
Cereal/Grain		
Oats	2.2	
Sorghum	1.5	
Wheat	2.8	0.6
Maize	5.0	0.7
Rice, paddy	4.2	0.4
Meat		
Chicken	0.6	0.3
Pork	0.2	
Beef	0.2	0.1

SOURCE: Jackie Hughes et al., "Vegetables for More Food Using Less Resources," World Vegetable Center, September 24, 2008, p. 2.

Using the same land area, you can obtain twelve times more food from vegetables than from legumes and five times more food than from cereals (table 12). If you factor in the amount of grain needed to produce meat, a single hectare of land can produce 29 times more food in the form of vegetables than in the form of chicken meat, 73 times more than pork, or 78 times more than beef. This makes vegetable production an ideal enterprise for smallholder farmers in developing countries, and for places where land is in short supply, such as cities. It is even suited to broadacre farming.[2]

Although vegetables need less land and other inputs, their cultivation is labor intensive, which ideally suits a world in which there is an abundance of people but other resources are becoming scarce. Typically, cultivating a field of vegetables employs from two to five times the number of people needed to cultivate the same area of grain. Vegetable culture offers worthwhile employment in both rural and urban areas, in poor communities and well-off ones alike—some of it extremely high-tech. Its expansion may help to stem the flow of rural people into the megacities, thus easing the pressure of the urban footprint on the surrounding countryside while restraining urban poverty and creating useful urban jobs.

The concept of the green city affords excellent opportunities for intensive vegetable culture on the roofs and walls and beneath buildings (using light pipes instead of electrical lighting) as well as on areas of interstitial land, in greenhouses, and in hydroponic and aquaponic facilities—though care must be taken to avoid the use of water and soil contaminated with sewage, heavy metals, chemicals, urban refuse, or pesticides. In periurban areas, where labor is plentiful, vegetable farming and associated activities along the market chain will enhance income and livelihoods. The World Vegetable Center says, "Vegetable production has a comparative advantage for urban and peri-urban areas, where arable land is scarce and labor is abundant as well as for rural areas where, given some basic rural infrastructure which permits access to markets, farming skills of rural families can be harnessed to alleviate poverty as well as to improve family and community nutrition."[3]

The power of vegetables to reduce poverty is significant: vegetables are more profitable to grow than rice or other grains, despite the fact that they require more labor. Their higher yields translate into a higher monetary return and so offer a faster route out of poverty for more people. The difference is the number of cropping days: as vegetables have a short growing cycle, many crops can be sown and harvested in a year, using the land more productively.[4]

Vegetable production is also more sparing in its use of water. The water needed to produce a kilogram (2.2 pounds) of beef, for example, can grow 91 kilograms (200 pounds) of tomatoes or 175 kilograms (385 pounds) of cabbages. The World Vegetable Center says that this makes vegetables particularly suitable for increasing food production in arid and semiarid regions, in cities, in places with declining water tables, or during drought.[5]

Vegetables can help overcome malnutrition: they generally contain higher levels of micronutrients such as vitamins A and C and folates, whereas cereals, legumes, and meat contain higher levels of macronutrients (carbohydrate, protein) and minerals such as iron. When the efficiency of land and water inputs into vegetable production is calculated, however, vegetables yield more energy, protein, vitamin A, and iron than meat per unit of land and water used. For the same area of land, vegetables yield slightly more iron than cereals, comparable levels of protein, but only half the energy yield. Legumes produce much higher energy, protein, and iron than vegetables per unit of cultivated land. Vegetables alone cannot compete with the energy provided by cereals and legumes, so these need to be in the diet, too. However, the lack of micronutrients in grains makes vegetable consumption important for a balanced and healthy diet. The World Vegetable Center argues that vegetables play a key role both in overcoming micronutrient malnutrition in the forms of undernutrition (starvation and malnourishment) and overnutrition (obesity and degenerative disease) and in alleviating poverty and improving gender equality, as women are the majority of the world's vegetable growers.

Most important, the world at large has yet to sample even a tiny proportion of the riches of the vegetable kingdom. Indigenous communities in Africa, for example, eat around four hundred kinds of vegetables quite unfamiliar to the Western palate. Similarly vast arrays of new flavors, tastes, and culinary experiences are available in India and Asia, Australasia, and Latin America. This is a culinary adventure on which the whole of humanity has yet to embark—and one that will make the sustaining of our food supply a great pleasure as well as a necessity.

Doubling the proportion of vegetables and fruits in the twenty-first-century global diet is an achievable goal, and not one that is in any way burdensome. This will necessarily involve reducing individual consumption of the high-energy foods such as meat, dairy, eggs, and fish in those societies where intake is already high, and avoiding high intakes in those where it is starting to climb. The fact that the Western diet kills more than half of its consumers through heart disease, cancer, stroke, and diabetes should be sufficient warning of its inherent risks—and reason enough for emerging economies to avoid its worst excesses and the phenomenal medical and health-care costs they entail. The twenty-first-century diet will be more healthful all round.

A Future for Meat

Eating more vegetables in the global diet should not affect the viability of the world's livestock producers, as world demand for protein overall will continue to rise with population and economic growth. Several factors are emerging, however, that are likely to transform contemporary livestock production: these include growing public resistance to factory farming and increasing public demand for clean, organically grown meat, the long-term cost of energy, the likely scarcity of feedgrains as global demand for food and fuel rises, the need to return carbon to the world's soil, and the increasing prevalence of drought under climate change.

All these are likely to encourage the second big-picture step we can take to solve the global food crisis, a trend back to the raising of high-quality livestock on the world's rangelands and savannahs—and a return to traditional low-input pastoralism of cattle, sheep, and goats. This time, however, it will be carried out with herd numbers strictly limited to what landscapes can sustain and using advanced technologies such as satellites and telemetry for remote management of animals, water, and pastures, with the ultimate goal of restoring fertility, vegetation, and carbon to the soil of the grasslands and deserts.

The need for more grazing will grow as the world's grain bowls dry out and cereal farming becomes more unreliable in these regions: a return to pastoralism is the best way to prevent this marginal land from becoming a weed-infested wasteland. In the future, meat may well be a high-priced luxury, consumed in small and delicate servings that reflect the true cost to the planet of producing it, and returning a better living to those who grow it than the industrialized and chemicalized bulk commodity of today does. There is no need for people to stop eating meat—merely to be more conservative in their consumption, more aware of the effects of eating meat on their world, and more willing to pay for prevention or repair of the damage.

So who is to encourage such a profound shift in human tastes and sustainable consciousness? The answer is the same people who helped us to overindulge in wasteful, destructive foods in the first place: the writers of cookbooks, the fast-food chains, the food industry, the supermarkets, the women's and lifestyle magazines, the dieticians and nutritional advisers, the agricultural bureaucracies, the food scientists. Maybe even celebrity TV chefs can help to save the planet.[6]

CURBING WASTE

Our second big-picture solution to the global food crisis is to cease wasting half the food we actually produce. Instead of wars on people or "terror," let's have a World War on Waste. If we can reduce the losses by even as much as half, we will feed two to three billion people well, besides using the world's water, energy, fertilizer, fish, and land resources more sparingly. And we will help prevent real wars.

Researchers at the Stockholm International Water Institute (SIWI) calculate that of the world's 2,700 cubic kilometers (675 cubic miles) of irrigation water, around half is wasted. Just the water wasted in the U.S. food system alone would grow enough food to feed five hundred million people, they note. "The amount of water that can be saved by reducing food waste is much larger than that saved by low-flush toilets and water-saving washing machines. It's time for us to move beyond thinking about how we meet quantities, and to start looking at the type of foods we produce and how we benefit from them," they argue. A large part of the solution lies in awareness: "Most urban consumers who were interviewed did not realise that meat, dairy and fruit come from living things that use natural resources to grow. With increased distance between farms and food consumption sites and commoditisation of food, the level of ignorance may only increase, and unaware consumers are less likely to question and change their behaviour," they warn.[7]

The SIWI team calls for a combination of policy measures, including price signals and scrutiny of food processors, outlets, and supermarkets, along with major public education campaigns starting in schools, to end the waste. The food chain is now very long and involves many actors between farmer and consumer, most of whom have little incentive to avoid wasting either food or water, they point out. This makes it imperative to engage everyone from food producer and trucker to business, government, and individual consumers and their children in the war on waste. Essential measures include the establishment of vast urban composting ventures that turn discarded food, organic factory waste, and garden and green clippings back into manure and fertilizer, stockfeed, compost, and soil for the production of food.

At the same time there must be an equally far-reaching effort to end the waste of water, land, and nutrients, in particular by the megacities. If water for food production runs short in certain areas, it may be necessary to quarantine rural supplies and to ration city water—or risk starvation. Likewise, it may be necessary to ban or restrict the development of land in the greater periurban catchment in order to preserve its food-producing and carbon-storage potential. Eventually, it will probably be necessary for all cities to recycle all of their water, their nutrients, and their waste—and the task of designing systems to do this that are safe, efficient, and wholesome is urgent.

Thrift has been a human virtue since the dawn of time. Virtually all our ancestors practiced it and would be horrified at our generation's neglect of this most basic tool for survival. As we approach the midcentury peak scarcity, it is time for today's societies and people to relearn and to attach new value to thrift. Wasting food, water, and nutrients makes no sense from anyone's point of view—farmer, family, manufacturer, supermarket, or government. Preventing it has many virtuous spinoffs, including stronger economic growth, new jobs, and new industries. Renewing our respect for thrift is a challenge for the world's religious, political, and community leaders.

SHARING KNOWLEDGE

The third big-picture way we can avoid the coming famine is through a major worldwide investment in new knowledge about sustainable ways to produce food, and by sharing new and existing knowledge far, far better among the world's 1.8 billion food producers. Norman Borlaug, one of the fathers of the Green Revolution, was a man who looked both back and far into the future. "In 1914, the year I was born, the world's population topped 1.6 billion," he wrote shortly before his death. "Today it is growing by more than 75 million per year. Since the middle of the last century, breakthroughs in agricultural science and technology have permitted global production of rice, wheat and other cereals to stay ahead of population growth, allowing millions to escape the pain of constant hunger. We know from long experience that advances in agricultural technology can buy time for political, social and religious leaders to bring into better balance the growth in human population and the carrying capacity of our planet."[8]

Given the complacency that has prevailed for more than thirty years, and the huge rise in human numbers and food demand of the recent past and of the coming decades, we should now be considering nothing less than a four- or fivefold increase in our investment in food and agricultural research and the communication of its outcomes with farmers worldwide. This should apply to all kinds of farming systems, affluent and poor, chemical and organic, large and small. It has to be for herders, pastoralists, and nomads as well as farmers, agroforesters, smallholders, aquaculturalists, and horticulturalists. Especially it has to be for the new generation of urban vegetable farmers. It should also be for food developers, dieticians, and cooks.

Knowledge usually delivers the greatest value to humanity when it is widely used. Investment and effort in knowledge sharing and dissemination should therefore equal that made in research and discovery and should exploit modern mass communication and educational tools, in order to reach every farmer and smallholder on Earth. The old model that handed scientists most of the money and left communicators practically nothing to disseminate new knowledge with has served humanity poorly, as it has failed to help many people and has, over time, led to the waste of a lot of good science. For the twenty-first century we need a new model that invests a dollar in knowledge sharing for every dollar in knowledge generation, and more if need be.

Although it is difficult to estimate exactly how much money is required to achieve global food security, raising the worldwide annual investment from $36 billion to $145 billion for both research and communication combined is a reasonable target, and also one that is readily affordable by the governments, food companies, and supermarkets of the world.[9] This level of investment would allow for a doubling in R&D and a matching investment in disseminating its outcomes to farmers and consumers. It is less than one-tenth of the annual global weapons budget.

In past history, rulers understood that food security was the bedrock of all subsequent governmental stability, economic growth, equity, and social progress. Without it, poverty, upheaval, and crisis are almost inevitable. Without food there is no stability, and without stability no government, education, health care, or civil society. Why so many nations have lost sight of this simple truth in recent times is hard to fathom. A perplexed World Bank observes,

> If agricultural growth has such unique abilities to reduce poverty, then why hasn't it been more consistently realized across developing countries? Poverty plummeted in China, India, Vietnam, and other countries when they went through major spurts of agricultural growth, just as industrial take-offs and rising incomes followed in the wake of major spurts of agricultural growth in Japan and the Republic of Korea. Yet agriculture has been used too little for growth and food security in today's agriculture-based countries, with high social costs. Its full abilities to reduce rural poverty have also been used too little in the transforming and urbanized developing countries, which have large populations of rural poor."[10]

In its 2008 World Development Report, the Bank mounted a persuasive case for top priority being given in policy to global food production.

There is a golden truth here that is often overlooked: poverty is seldom, if ever, solved by throwing money at it. The Grameen Bank founder and Nobel peace laureate Muhammad Yunus says, of microcredit, "The first and foremost task . . . is to turn on the engine of creativity inside each person."[11] The same can be achieved by sharing knowledge. Knowledge gives to all people the power, the opportunity, and the confidence to better themselves, to be more productive, to educate their children, to secure their future, and to shape their own destinies. It offers a fresh sunrise, a new horizon. The first and most important knowledge is how to feed oneself sustainably; delivering it to every citizen on Earth who wants it is neither very difficult nor very expensive, compared to other human enterprises.

Knowledge about how to produce food sustainably is, and always should be, free to all people. This is a moral principle for our time, the century of scarcities. It ought not to be owned and withheld by corporations, institutions, or countries.

A second moral principle is that those who profit most from food—the manufacturers and supermarkets—should volunteer, or be required, to devote a slice of their profits to the agricultural science and communication necessary to secure the future food supply. This is, after all, little more than enlightened self-interest on their part. And since governments and farmers have been working so hard for so long to "add value" to food so that the food companies can make greater profits, these corporations ought to be willing to contribute a fair percentage of their gain to research and dissemination of sustainable agricultural know-how. Just as mining companies are now asked to clean up the sites they mine and repair their environmental damage, so should the food industry be required to repair the environmental damage it causes through its pricing policies. Indeed, contributing to agricultural research and knowledge sharing should be a condition of operating as a food processor, wholesaler, or retailer—and those who cannot truthfully account for this financial support on their labels and in their advertising should be shunned by consumers.

Food policy, as we have seen, is also defense policy. Countries that do not want their borders stormed by tens of millions of starving immigrants and refugees need to ask themselves: what is the best way to avoid this—to prevent starvation in the first place, or to commit two acts of absolute indecency by permitting starvation, and then employing military and police force against the starving? Countries that do not want their children caught up in other people's, potentially nuclear, conflicts need to focus on the most effective means to allay the tensions. Once it becomes clear to governments that food scarcity is a major source of tension and an instigator of wars, the logic of investing part of the defense budget in war prevention through food security becomes clear. It ceases to be "charity" and starts to be viewed as self-preservation.

If the price of world peace is to divert just one-tenth of the global armaments spending of US$1.5 trillion to sustaining the food supply, it is surely not too high a price for any rationally led or ethical country to afford. The funds to achieve this exist and are already being invested in protecting every nation—but mainly in weapons of destruction rather than instruments of peace, creation, and renewal. Unfortunately, weapons of destruction get used sooner or later. So it is not a case of stressed national budgets being asked to find new funding for food security but rather of reallocating defense budgets that already exist to war prevention through a sustainable food supply and the easing of internal and international tensions. Interestingly, soldiers appear to understand this far better than politicians do.

RECARBONIZING

Climate change is the shadow on the human destiny. Its most devastating impact will be felt in food insecurity and the problems that flow from it. Addressing it is the fourth big-picture priority for avoiding the coming famine.

It may already be too late to prevent the sort of warming that will disrupt agriculture worldwide by the middle of the century, so the immediate food strategy must be one of adaptation to uncertain climates. This again demands massive investment in agricultural science and knowledge sharing, focused on the needs of both broadacre and smallholder farmers. We need urgently to develop hardier varieties of the world's staple food crops and pastures and to deliver them to farmers everywhere, as swiftly as possible. We need to scour the planetary gene pool for novel crops or genes that can sustain the yield of food when the rains fail, the heat rises, or the floods and pests come.

Overcoming the impact of climate change on food security also necessitates the demolition of global trade barriers and subsidies, so that food may be produced efficiently in the most reliable places and exported economically to those facing shortages, and so that virtual water and virtual energy may develop real instead of theoretical meaning. Just as food policy is defense policy, it is also trade policy—and the nation that forgets this as we approach the midcentury peak of demand and scarcity increases its vulnerability to conflict and migratory crisis.

There is also much that agriculture, agroforestry, and forestry can do to help limit global warming; to exclude them from climate policies is asking for trouble. As we have seen, they generate up to 30 percent of global greenhouse emissions, and these must be reduced by all possible means—including modifying our diet so we "eat" fewer barrels of oil. An equally urgent goal is to recarbonize our depleted soil, thus restoring its fertility, by adding carbon or organic matter through well-designed farming and grazing systems that capture these. We also need to discover ways to manage the microscopic life within the soil so well that it releases nutrients we cannot now access, helps to protect crops, and increases their yields. This eco-farming is, potentially, the foundation of the next agricultural revolution, but much public-sector research remains to be done to understand and develop it. By locking up more carbon in soil, trees, and pastures, agriculture can help significantly to draw down the amount of carbon in the atmosphere and curb the impetus for climate change.

TOWARD THE WORLD FARM

Beyond a doubt, the twenty-first century will be the most decisive ever faced by humanity, and the coming famine may be the instrument by which civilization's long forward and upward march over four millennia is most likely to be disrupted, even thrown into reverse. This is the period in which humanity either attains the maturity and cooperative spirit to overcome another big challenge to its survival—or bows to the hunger, violence, and systemic failure of a new Dark Age.

Our dominance of the Earth's systems—atmosphere, oceans, freshwater, soil, nutrient and energy flows, and biodiversity—is now so overwhelming that we can no longer depend on these elemental forces to function or to regenerate on their own, naturally and without our involvement. We are in a position where, in order to survive ourselves, we need to take knowledgeable responsibility for the changes we have wrought on a planetary scale and to manage the conditions we have created in as stable a fashion as we can. In the case of climate change, we are already recognizing that, having mismanaged the atmosphere, we need to start managing it better. In the case of the oceans and savannahs, although we recognize the mismanagement we have yet to accept full responsibility for the mess in terms of repairing it. In the case of the forests, it's a bit of both.

Climate change may already have progressed so far that, to prevent the planet from overheating to intolerable levels, humanity may have scant alternative than to intervene in the entire atmosphere, much as we manage the air in an enormous building. Scientists are already debating a range of astonishing interventions such as the injection of sulfur dioxide particles into the upper atmosphere or the use of space-borne mirrors to reflect sunlight away from the Earth, as devices to bring about temporary global cooling and buy time as we struggle to agree on ways to cut greenhouse emissions. These are desperate ideas indeed as, like the emission of carbon dioxide itself, they carry with them the risk of incalculable side effects. But they underline an inescapable point: humanity is wholly and solely in charge of its own fate now; there's no one else to blame or to turn to.

If we are considering ways to manage something as vast and complex as the atmosphere, we should be prepared to consider a global approach to managing the *trophosphere*. The word is a neologism (not to be confused with *troposphere,* meaning the lower atmosphere) coined to describe the entire Earth system that nourishes us: soil, water, nutrients, energy flows, marine life, plants, animals, insects, soil biota, and so on. Hitherto we have managed these as discrete individual, local, regional, biological, and national issues, industries, and activities. The time is fast approaching when we need to think of the food system as a single, planetary whole, so it may sustain us through the midcentury clash of needs with scarcities. Radical though such an idea may appear in the context of a world fragmented by national, ideological, religious, cultural, ethical, technological, and other differences, it is no more than a logical extension of all that has gone on in the past two centuries, as farms and rural industries grew larger, integrated, and amalgamated; knowledge and technology were more widely distributed; farmers cooperated on water or soil management; governments took greater oversight of health, safety, and sustainability; and food standards increased. These things have been achieved at local, regional, and national levels through science, education, consensus, and regulation, and the next step is to apply them planetwide.

For this to work, governments need not only to agree on common policies for the allocation, use, and management of water, land, nutrients, and other inputs, but also to enlist the willing participation of farmers, agribusiness companies, supermarkets, and everyone in the distributional chain—consumers in particular, as they are the ones who send out the vital market signals. An essential first step is the abolition of trade barriers and subsidies, so all farmers receive the same signals and agriculture can move to where it is most advantageous and secure to grow food. This is no longer a matter of trade talks and nationalistic grandstanding. It is a matter of survival. The countries that oppose it must know that they are fostering starvation, crisis, and war—and be condemned as such by others.

A second step is to develop and rapidly share some universal best-practice ideas for looking after water and land, for low-energy sustainable farming, for reducing waste in the food chain, for establishing and promoting low-energy diets, for recycling urban waste into food-producing systems novel and traditional, and for establishing green cities. The world farm, a venture consisting of 1.8 billion individual (and highly individualistic) producers, is never going to function like an industrial corporation—nor should it. Its strength lies in its rich diversity. But in an age of scarcity, feeding humanity sustainably is a single enterprise and should come to be viewed as one, operating, as farmers mostly do, on goodwill, consensus, knowledge-sharing, and cooperation. It should be run by the people who know how to do this best, the farmers—with all the help we can possibly give them in the form of knowledge, technology, good governance, fair prices, environmental assistance, and the moderation of our demands on their basic resources.

A third essential step is to find a way for supermarkets, the food industry, and governments to pay farmers not only what it takes to produce high-quality food—but also to look after the environment and production system for the longer term. At present the world's farmers are paid in a fashion that exploits both them and the planet, discourages investment in new, more sustainable technologies, and deters young people from taking up agriculture. People have no problem with the idea of fair pay for bankers, lawyers, unionists, nurses, actors, public servants, economists, or politicians—so what is wrong with fair pay for farmers? This is not mere sophistry: it is about paying farmers a price that enables them to husband the food-producing capacity of the Earth. It is about understanding that when we underpay farmers, we risk our own future.[12]

Just as users of fossil fuels are increasingly being asked to pay for the real cost of those fuels to the global climate and for the switch to alternative energies through carbon pricing, consumers will sooner or later have to shoulder the real cost of food to the environment and society—not an artificially low price that damages these things. The time has come for the food processing and retail industry to accept responsibility for the harm it now causes through unrealistically low prices that undermine farmers and Earth systems—and to pass that cost on to its customers. To help them do this, however, we can educate the coming generation of the world's children about food, and what it really takes to produce it.

Today's food is too cheap to last. To avert the coming famine we all need to start paying its true price—not blindly transferring the cost of what we consume today to our grandchildren tomorrow.

A CALL TO ACTION

Like the global climate to which we are all subject, the global food supply will decree our fate, no matter who we are or where we live. If through our neglect or abuse of resources it fails, each of us will bear the consequences.

Humans are wiser than that, however. In the twentieth century we gained the means to wreck civilization with fearsome weapons—and somehow at the same time acquired the wisdom not to do so. We tore a hole in the ozone layer—and then agreed to repair it. We have lately come to appreciate our ability to destroy the very climate that birthed our civilization—and we are embarking on the first collaborative steps to remedy it. Each situation demanded a greater effort in societal and international cooperation, education, personal commitment, public-private alliances, and the burying of age-old differences, hatreds, and fears.

We are now just starting to appreciate how our numbers, our appetites, and our profligate use of resources have combined to place us in a jeopardy as great as any of these still-existing perils. Yet the answers are the same: cooperation, sharing of knowledge, and the subordination of national pride, greed, and fear will see us through.

This time, however, there is a larger role for the individual, who must face personal choices about both their protein intake and their fertility.

Each danger we confront and successfully surmount takes us a step closer to becoming a single people on a single planet, differentiated though we may choose to remain as individuals or cultures. The coming famine is the trial of our common humanity.

For centuries we have gathered over the meal table to enjoy the bounteous gifts of nature and our farms, as well as one another's good company and hospitality. Food has always helped us sort out our differences and make new friends. Working together across a world to secure our food supply offers us a fresh opportunity to appreciate the limitless virtues and qualities of our fellow humans—and how wonderful and powerful these things can be when we braid them together and allow them to draw us far into our common future.

Conversions

The following conversions may assist in comparing the different sources cited in the book.

1 liter water	weighs 1 kilogram (2.2 lbs)
10 liters	2.64 U.S. gallons
1 kiloliter	1,000 liters or 1 cubic meter of water or 264 U.S. gallons
1 cubic meter of water	1,000 liters, which weighs 1 tonne (1,000 kilograms)
1 megaliter	1,000,000 liters of water, which weighs 1,000 tonnes
1 Olympic pool	2.5 megaliters of water
1 gigaliter	1,000,000,000 liters of water
1 teraliter	1,000,000,000,000 liters or 1 cubic kilometer
1 cubic kilometer	1,000,000,000,000 liters or 1 teraliter or 264 billion U.S. gallons
1 hectare	10,000 square meters or 2.47 acres
1 square kilometer	1,000,000 square meters or 100 hectares or 247 acres

1 kilogram	2.2 lbs
1 tonne	1,000 kilograms or 1.1 U.S. (short) tons
1 calorie (nutritional)	4.186 kilojoules

Notes

1. WHAT FOOD CRISIS?

1. "G8 Leaders Statement on Global Food Security," G8 Summits Hokkaido Official Documents, July 8, 2008, www.g8.utoronto.ca/summit/2008hokkaido/2008-food.html.

2. Ibid.

3. Andrew Grice, "Over Caviar and Sea Urchin, G8 Leaders Mull Food Crisis," *London Independent,* July 8, 2008, www.independent.co.uk/news/world/politics/over-caviar-and-sea-urchin-g8-leaders-mull-food-crisis-862051.html.

4. "Global Food Crisis 'Silent Tsunami' Threatening over 100 Million People, Warns UN," UN News Centre, April 22, 2008, www.un.org/apps/news/story.asp?NewsID=26412&Cr=food&Cr1=price. These figures assume a cost of eighty cents per day to feed each person.

5. "Higher Grain Yield Seen Needed for Food/Fuel Supply," Reuters, April 17, 2007, www.reuters.com/article/environmentNews/idUSL1750757220070417; Lester R. Brown, "World Facing Huge New Challenge on Food Front: Business-as-Usual Not a Viable Option," Earth Policy Institute, April 16, 2008, www.earth-policy.org/Updates/2008/Update72.htm.

6. Lee Glendinning, "Overweight Now Outnumber Underfed around the World," Times (London) Online, April 15, 2006, www.timesonline.co.uk/tol/news/world/article609571.ece. According to the Global Health Council, more than twenty-eight thousand children die each day, 53 percent from hunger-related causes. See Global Health Council, "Child Mortality," n.d., www.globalhealth.org/child_health/child_mortality/.

7. Anthony Faiola, "The New Economics of Hunger," *Washington Post,* April 27, 2008, www.washingtonpost.com/wp-dyn/content/story/2008/04/26/ST2008042602333.html.

8. Vivienne Walt, "The World's Growing Food-Price Crisis," *Time,* February 27, 2008, www.time.com/time/world/article/0,8599,1717572,00.html.

9. Louise O. Fresco, "Biomass, Food and Sustainability—Is There a Dilemma?" Rabobank, autumn 2007, www.rabobank.com/content/images/Biomass_food_and_sustainability_tcm43–38549.pdf; Benjamin Senauer, "The Appetite for Biofuel Starves the Poor," *Guardian,* July 3, 2008, www.guardian.co.uk/commentisfree/2008/jul/03/biofuels.usa; "Man-Made Hunger," editorial, *New York Times,* July 6, 2008, www.nytimes.com/2008/07/06/opinion/06sun1.html?scp=2&sq=food+riots+30+countries&st–yt; "World Bank President to G8: 'World Entering a Danger Zone,'" World Bank, press release, July 2, 2008, http://web.worldbank.org/WBSITE/EXTERNAL/NEWS/0,,contentMDK:21828803~pagePK:34370~piPK:34424~theSitePK:4607,00.html; "Countries in Crisis Requiring External Assistance," *Crop Prospects and Food Situation* no. 2 (April 2008), www.fao.org/docrep/010/ai465e/ai465e02.htm; "Food Prices—The Silent Tsunami," *Economist,* April 19, 2008, www.economist.com/opinion/displayStory.cfm?Story_ID=11050146.

10. Karen Barlow, "Food Riots an Apocalyptic Warning," ABC News (Australia), April 14, 2008, www.abc.net.au/news/stories/2008/04/14/2215873.htm?section=world.

11. "World Bank Chief Calls for G8 Help to Feed World's Hungry," Environment News Service, July 8, 2008, www.ens-newswire.com/ens/jul2008/2008–07–03–01.asp; David Nason, "First Signs of the Coming Famine," *Australian,* April 26, 2008, www.theaustralian.news.com.au/story/0,25197,23599517–27702,00.html.

12. "High Rice Prices No Windfall for Many Asian Farmers," International Rice Research Institute, April 7, 2008, http://beta.irri.org/news/index2.php?option=com_content&do_pdf=1&id=3928; "Major Fast-Food Companies to Serve Half-Cup Rice," *Manila Times,* April 2, 2008, p. 1, www.manilatimes.net/national/2008/apr/02/yehey/images/front.pdf; "Rice Heist Sows Panic among Farmers," Reuters, March 24, 2008, www.reuters.com/article/oddlyEnoughNews/idUSBKK31600200803 24; Suzy Jagger and David Charter, Dow-Jones News Wire, "Wal-Mart, UK Stores Ration Rice Sales," *Australian,* April 24, 2008, www.theaustralian.news.com.au/story/0,25197,23591506–23850,00.html; "Factbox: Why Have Rice Prices Surged to Record Highs?" Reuters, April 25, 2008, www.reuters.com/article/worldNews/idUSSP13081120080425?pageNumber=2&virtualBrandChannel=0&sp=true; "Rice Prices 'to Keep on Rising,'" BBC, April 11, 2008, http://news.bbc.co.uk/2/hi/business/7341978.stm.

13. International Food Policy Institute, "New Global Hunger Index: 33 Countries Have 'Alarming' or 'Extremely Alarming' Levels of Hunger," press release, October 15, 2008, www.ifpri.org/PRESSREL/2008/pressrel20081014.pdf; Food and Agriculture Organization of the United Nations, "Bill Gates, FAO Chief Ponder Anti-hunger Fight," FAO Newsroom, May 11, 2009, www.fao.org/news/story/en/item/19516/icode/.

14. Joachim von Braun et al., "Overview of the World Food Situation," brief prepared for the Annual General Meeting of the Consultative Group on International Agricultural Research, Nairobi, October 29, 2003, www.ifpri.org/pubs/speeches/20031029vonbraun.htm; Organization for Economic Coopera-

tion and Development and Food and Agriculture Organization of the United Nations, "OECD-FAO Agricultural Outlook 2008–2017," 2008, p. 11, www.oecd.org/dataoecd/54/15/40715381.pdf.

15. Aditya Chakrabortyy, "Secret Report: Biofuels Caused Food Crisis," *Guardian,* July 3, 2008, www.guardian.co.uk/environment/2008/jul/03/biofuels.renewableenergy; Joachim von Braun, "High and Rising Food Prices: Why Are They Rising, Who Is Affected, How Are They Affected, and What Should Be Done?" paper presented at U.S. Agency for International Development conference, Washington, D.C., April 11, 2008, p. 4, www.ifpri.org/presentations/20080411jvbfoodprices.pdf.

16. Faiola, "New Economics of Hunger"; "Philippines Caught in Rice Squeeze," *Washington Post,* April 12, 2008, p.1, www.washingtonpost.com/wp-dyn/content/article/2008/04/11/AR2008041104162.html.

17. For example, John Vidal, "Global Food Crisis Looms as Climate Change and Fuel Shortages Bite," *The Guardian,* November 3, 2007, www.guardian.co.uk/environment/2007/nov/03/food.climatechange; John Iroga, "Climate Change Linked to Global Food Shortage," *Solomon Times,* May 1, 2008, www.solomontimes.com/news.aspx?nwID=1705; "Climate Change May Intensify Food Crisis, Warns UN," *Sindh Today,* February 17, 2009, www.sindhtoday.net/world/64875.htm.

18. Andrew Stoekel, "High Food Prices: Causes, Implications and Solutions," paper prepared for the Rural Industries Research and Development Corporation, Government of Australia, Publication 08–1900, June 2008, https://rirdc.infoservices.com.au/downloads/08–100.pdf.

19. Bruce Sundquist, "The Global Food Crisis: Some Solutions for a World with Fewer Options for Satisfying Increasing Demands," September 2008, http://home.windstream.net/bsundquist1/index.html.

20. Michael Murphy, "The Era of Cheap Food Is Well and Truly Over!" DairyNZ, June 2008, p. 10, www.dairynz.co.nz/file/fileid/8947.

21. Dow Jones Newswire, "Global Food Demand to Double by 2050, Research Key: Experts," Flexnews: Business News for the Food Industry, October 16, 2007, www.flex-news-food.com/pages/11603/global_food_demand_double_2050_research_key___experts_dj.html; FAO, Declaration of the World Summit on Food Security, Rome, November 16–18, 2009, p. 2, http://www.fao.org/fileadmin/templates/wsfs/Summit/Docs/Final_Declaration/WSFS09_Declaration.pdf.

22. International Water Management Institute, summary of *Water for Food; Water for Life: A Comprehensive Assessment of Water Management in Agriculture,* ed. David Molden (Sterling, Va.: Earthscan, 2007), p. 5, www.fao.org/nr/water/docs/Summary_SynthesisBook.pdf. I based my projection of growth in urban demand on United Nations population projections.

23. Ron Leng, "The Morality of Biofuels," editorial, *ScienceAlert,* May 7, 2007, www.sciencealert.com.au/opinions/20070805–15733.html.

24. Boris Worm et al., "Impacts of Biodiversity Loss on Ocean Ecosystem Services," *Science* 314 (November 3, 2006): 789.

25. Alex Evans, "The Feeding of Nine Billion: Global Food Security for the 21st Century, a Chatham House Report," January 26, 2009, p. 6, www

.humansecuritygateway.info/documents/CHATHAM_FeedingNineBillion_GlobalFoodSecurity21stCentury.pdf.

26. Ibid.

2. FOOD . . . OR WAR?

1. Jimmy Carter, "First Step to Peace Is Eradicating Hunger," *International Herald Tribune,* June 17, 1999, www.iht.com/articles/1999/06/17/edcarter.2.t.php.

2. Richard Pipes, *A Concise History of the Russian Revolution* (New York: Alfred A. Knopf, 1995).

3. Carter, "First Step to Peace Is Eradicating Hunger."

4. The International Food Policy Research Institute, for example, defines "food wars" specifically as the use of hunger as a weapon or hunger that follows from destructive conflict, not vice versa. The Food and Agriculture Organization of the United Nations (FAO), in many papers, attributes food scarcity to wars—but almost no wars to food scarcity. See, for example, FAO, "World Food Summit Five Years Later: Conflict and Food Insecurity," FAO, 2007, www.fao.org/worldfoodsummit/msd/Y6808e.htm.

5. Indra de Soysa and Nils Petter Gleditsch with Michael Gibson, Margareta Sollenberg, and Arthur H. Westing, "To Cultivate Peace: Agriculture in a World of Conflict," PRIO Report 1/99, 1999, www.isn.ethz.ch/isn/Digital-Library/Publications/Detail/?ots591=CAB359A3–9328–19CC-A1D2–8023E646B22C&lng=en&id=37962.

6. Ibid.

7. Ibid.

8. In 2006 the countries with the highest rates of mortality for children younger than five were Sierra Leone (270 per 1,000), Angola (260 per 1,000), Afghanistan (257 per 1,000), Niger (253 per 1,000), Liberia (235 per 1,000), Mali (217 per 1,000), Chad (209 per 1,000), Equatorial Guinea (206 per 1,000), Democratic Republic of the Congo (205 per 1,000), Burkina Faso (204 per 1,000), and Guinea-Bissau (200 per 1,000). See UNICEF, WHO, World Bank, and UN Population Division, "Levels and Trends of Child Mortality in 2006: Estimates Developed by the Inter-agency Group for Child Mortality Estimation," New York, 2007, p. 29, www.unescap.org/rw-mdgm/REG/Indicator4.1(Document1).pdf.

9. Ellen Messer and Mark J. Cohen, "Conflict and Food Security," International Food Policy Research Institute, August 2001, www.ifpri.org/2020/focus/focus08/focus08_12.asp.

10. "Poverty 'Fuelling Terrorism,'" BBC News, March 22, 2002, http://news.bbc.co.uk/1/hi/world/1886617.stm.

11. Ben Sutherland, "Water Shortages 'Foster Terrorism,'" BBC News, March 18, 2003, http://news.bbc.co.uk/2/hi/science/nature/2859937.stm.

12. Ellen Messer, Mark J. Cohen, and Jashinta D'Costa, "Food from Peace: Breaking the Links between Conflict and Hunger," International Food Policy

Research Institute, 2020 brief no. 50, June 1998, www.ifpri.org/2020/BRIEFS/NUMBER50.HTM.

13. FAO, "Assessment of the World Food Security and Nutrition Situation," report to the Committee on World Food Security, 34th sess., October 14–17, 2008, Rome, www.fao.org/UNFAO/Bodies/cfs/cfs34/index_en.htm.

14. U.S. Central Intelligence Agency, "Transnational Issues," *World Fact Book,* April 29, 2009, https://www.cia.gov/library/publications/the-world-factbook/geos/xx.html.

15. Alan Dupont, *East Asia Imperilled: Transnational Challenges to Security* (Cambridge: Cambridge University Press, 2001), pp. 109, 110; Lee Hsieng Loong, keynote address, International Institute for Strategic Studies 7th Asia Security Summit Shangri-la Dialogue, May 30, 2008, Singapore, www.iiss.org/conferences/the-shangri-la-dialogue/shangri-la-dialogue-2008/plenary-session-speeches-2008/keynote-address-lee-hsien-loong/?locale=en.

16. David McKeeby, "Many Modern Conflicts Are Food Wars, Say Experts," America.gov (a Web site of the U.S. State Department), September 12, 2007, www.america.gov/st/washfile-english/2007/September/20070912113737idybeekcmo.9691278.html.

17. "Red Cross Warns of New Food Riots," virginmedia.com, May 27, 2008, http://latestnews.virginmedia.com/news/world/2008/05/27/red_cross_warns_of_new_food_riots.

18. Ismail Serageldin, "Avoiding Water Wars: A Vision for a Water Secure World," Barcelona, 2007, www.serageldin.com/ListPresentations.aspx; Mike Thomson, "Ex-UN Chief Warns of Water Wars," BBC News, February 2, 2005, http://news.bbc.co.uk/2/hi/africa/4227869.stm; Ban Ki-moon, "Time Is Running Out on Water," remarks prepared for delivery to World Economic Summit, Davos, January 24, 2008, www.un.org/apps/sg/sgstats.asp?nid=2967.

19. Adel Darwish, "Analysis: Middle East Water Wars," BBC News, May 30, 2003, http://news.bbc.co.uk/2/hi/middle_east/2949768.stm.

20. Peter H. Gleick, "Water Conflict Chronology," Pacific Institute for Studies in Development, Environment, and Security, November 10, 2008, www.worldwater.org/conflictchronology.pdf.

21. Dupont, *East Asia Imperilled.*

22. Shira B. Yoffe, "Basins at Risk: Conflict and Cooperation over International Freshwater Resources: Indicators and Findings of the Basins at Risk Project" (Ph.D. diss., October 2001, Oregon State University), www.transboundarywaters.orst.edu/research/basins_at_risk/index.html.

23. Gleick, "Water Conflict Chronology."

24. Dupont, *East Asia Imperilled,* p. 101.

25. Ibid., p. 109.

26. Meryl J. Williams and Choo Poh Sze, "Fish Wars: Science Is Shaping a New Peace Agenda," p. 85, paper prepared for "Food, Water and War: Security in a World of Conflict," conference sponsored by the Crawford Fund for International Agricultural Research, Canberra, August 15, 2000.

27. De Soysa, Gleditsch, et al., "To Cultivate Peace."

28. United Nations High Commissioner for Refugees, "2008 Global Trends: Refugees, Asylum-Seekers, Returnees, Internally-Displaced and Stateless Persons," June 2009, http://www.unhcr.org/4a375c426.html.

29. Article 1A(2), Convention and Protocol Relating to the Status of Refugees, 1951 and 1967, www.unhcr.org/protect/PROTECTION/3b66c2aa10.pdf.

30. For Canadian immigration statistics, see http://epe.lac-bac.gc.ca/100/202/301/immigration_statistics-ef/index.html; statistics regarding legal permanent residents of the United States are available at www.dhs.gov/ximgtn/statistics/publications/yearbook.shtm.

31. Steven Mintz, "The Irish Potato Famine," *Digital History,* 2007, www.digitalhistory.uh.edu/historyonline/irish_potato_famine.cfm.

32. Development, Concepts and Doctrine Centre, *The DCDC Global Strategic Trends Programme, 2007–2036* (London: U.K. Ministry of Defence, 2007), pp. 27–31, www.dcdc-strategictrends.org.uk/viewdoc.aspx?doc=1.

33. Kurt M. Campbell et al., "The Age of Consequences: The Foreign Policy and National Security Implications of Global Climate Change," Center for Strategic and International Studies and Center for a New American Security, Washington, D.C., November 2007, pp. 6, 7, www.csis.org/media/csis/pubs/071105_ageofconsequences.pdf.

34. Sam Perlo-Freeman, Catalina Perdomo, Elisabeth Sköns, and Petter Stålenheim, summary of chap. 5, "Military Expenditure," in Stockholm International Peace Research Institute, *SIPRI Yearbook 2009: Armaments, Disarmament and International Security,* www.sipri.org/yearbook/2009/05.

3. THE WELL RUNS DRY

1. International Riverfoundation, "The Dying Lake," 2006 brochure, Brisbane. See also Amos Brandeis, "Lake Bam—Twinning Project between Alexander River Restoration Administration and Burkina Faso, Africa," RestorationPlanning.com, n.d., www.restorationplanning.com/bam.html.

2. Associated Press, "Shrinking of Lake Chad: A Tale of Human Abuse, Climate Change," Global Policy Forum, December 14, 2006, www.globalpolicy.org/security/natres/water/2006/1214lakechad.htm; Food and Agriculture Organization of the United Nations (FAO), "Lake Chad Facing Humanitarian Disaster," October 15, 2009, www.fao.org/news/story/en/item/36126/icode/; "China's Water Shortage to Hit Danger Limit in 2030," People's Daily On Line, November 16, 2001, http://english.peopledaily.com.cn/200111/16/eng20011116_84668.shtml.

3. International Water Management Institute (IWMI), "Water: The Forgotten Crisis," press release, July 10, 2008, www.iwmi.cgiar.org/news_room/pdf/Water_The_forgotten_crisis.pdf.

4. David Molden, "Achieving Food and Environmental Security: Better Water Management for Healthy Coastal Zones," keynote address for Environment and Livelihoods in Coastal Zones, Bac Lieu, Vietnam, March 1–3, 2005.

5. IWMI, summary of *Water for Food; Water for Life: A Comprehensive Assessment of Water Management in Agriculture,* ed. David Molden (Sterling,

Va.: Earthscan, 2007), www.fao.org/nr/water/docs/Summary_SynthesisBook.pdf.

6. IWMI, "Water Management: Urgent Need for More Food with Less Water," *Science Daily,* March 27, 2007, www.iwmi.cgiar.org/News_Room/Multimedia/coverage/pdf/ScienceDaily_Urgent_need_for_more_food_with_less_water.pdf; FAO, *FAO Statistical Yearbook 2005–2006,* table D.1, www.fao.org/es/ess/yearbook/vol_1_1/pdf/d01.pdf.

7. Wayne S. Meyer, "Water for Food—The Continuing Debate," Land and Water, Commonwealth Scientific and Industrial Research Organisation, Government of Australia, n.d., www.clw.csiro.au/issues/water/water_for_food.html#howmuch.

8. Similarly, a motorist filling his or her vehicle's fuel tank with biodiesel will in fact be "fueling" that car with 30 tonnes of water (33 U.S. tons), on average once a week. More on this in chapter 8.

9. IWMI, summary of *Water for Food,* p. 5.

10. A.K. Chapagain and A.Y. Hoekstra, "Water Footprint of Nations," table 4.9, vol. 1, Main Report, Value of Water Report no. 16, UNESCO Institute for Water Education, Netherlands, November 2004, www.waterfootprint.org/Reports/Report16Vol1.pdf. I have converted the figures from cubic meters (as given in the report) to tonnes (1 cubic meter=1 tonne).

11. See IWMI, summary of *Water for Food,* pp. 7 and 14, on the doubling of the area of irrigated land and the tripling of water withdrawals. The estimate of freshwater needs comes from Colin Chartres, "Only a Price on Water Can End Threat to Food Security," *ScienceAlert,* August 21, 2009, www.sciencealert.com.au/opinions/20092108–19601.html.

12. Extrapolation of UN estimate that cities housed 37 percent of global population in 1975 and 47 percent in 2000, and will have 53 percent in 2015 and 60 percent in 2030 (see Population Division of the Department of Economic and Social Affairs of the United Nations Secretariat, *World Population Prospects: The 2006 Revision* and *World Urbanization Prospects: The 2007 Revision,* http://esa.un.org/unup). Intensification of food production could shift a further 1 billion or so people to the cities from rural areas. The figure of 1,200 cubic kilometers comes from IWMI, summary of *Water for Food,* p. 6, and the figure of 2,500 cubic kilometers is my projection.

13. David Krantz and Brad Kifferstein, "Water Pollution and Society," n.d., www.umich.edu/~gs265/society/waterpollution.htm; Peter H. Gleick, "Dirty Water: Estimated Deaths from Water-Related Disease 2000–2020," Pacific Institute Research Report, August 15, 2002, p. 3, www.pacinst.org/reports/water_related_deaths/water_related_deaths_report.pdf.

14. IWMI, summary of *Water for Food,* p. 1.

15. David Pimentel et al., "Will Limits of the Earth's Resources Control Human Numbers?" paper, College of Agriculture and Life Sciences, Cornell University, Ithaca, N.Y., February 25, 1999, http://dieoff.org/page174.htm.

16. Joachim von Braun et al., "Overview of the World Food Situation," brief prepared for the annual general meeting of the Consultative Group on International Agricultural Research, Nairobi, October 29, 2003, www.ifpri.org/pubs/speeches/20031029vonbraun.htm.

17. V. Haller, "Water an Issue at Food Summit," *Jakarta Post,* November 11, 1996.

18. "China's Water Shortage."

19. Quotations from John McAlister, "China's Water Crisis—Strategy Focus," Deutsche Bank China Expert Series, March 22, 2005; Ruixiang Zhu, "China's South-North Water Transfer Project and Its Impacts on Economic and Social Development," Ministry of Water Resources, People's Republic of China, n.d., www.mwr.gov.cn/english1/20060110/20060110104100XDENTE.pdf.

20. IWMI, "Water: The Forgotten Crisis"; Chartres, "Only a Price on Water Can End Threat to Food Security."

21. Brian Halweil, "Meat Production Continues to Rise," Worldwatch Institute, August 20, 2008, www.worldwatch.org/node/5443?emc=el&m=136135&l=5&v=ea5c798974; McAlister, "China's Water Crisis"; Lenntech, "Use of Water in Food and Agriculture," Lenntech.com, n.d., www.lenntech.com/water-food-agriculture.htm. The amount of water required to meet the increase in meat demand is my extrapolation from figures provided by the sources cited. This estimate of meat demand may also be too low because, with the predicted collapse in global fisheries, it may be necessary to find an additional 100 million tonnes (110 million U.S. tons) of meat or other protein from land-based sources. See chapter 6.

22. IWMI, "Water: The Forgotten Crisis."

23. Some scientists even argue that part of the Earth's groundwater percolates as superheated steam out of the planet's liquid mantle and is thus replaced only at geological rates over millions of years. See, for example, Lance Endersbee, *A Voyage of Discovery* (Frankston, Victoria, Aus.: Author, 2005).

24. "How Much Do We Depend on Groundwater?" Groundwater Foundation, n.d., www.groundwater.org/gi/depend.html; David E. Kromm, "Ogallala Aquifer," *Water Encyclopedia,* n.d., www.waterencyclopedia.com/Oc-Po/Ogallala-Aquifer.html; Manjula V. Guru and James E. Horne, "The Ogallala Aquifer," Kerr Center for Sustainable Agriculture, July 20, www.kerrcenter.com/publications/ogallala_aquifer.pdf; J. J. Woods et al., "Water Level Decline in the Ogallala Aquifer," Kansas Geological Survey Open-file Report 2000–29B (v2.0), December 1, 2000, www.kgs.ku.edu/HighPlains/2000–29B/Decdir.htm; Docking Institute of Public Affairs, "The Value of Ogallala Aquifer Water in Southwest Kansas," figs. 2 and 3 in chap. 1 of *The Value of Ogallala Groundwater* (Hays, Kans.: Docking Institute, 2001), www.fhsu.edu/docking/img/Archives/SW%20Groundwater-Ogallala/Part%204.Chapter%201.pdf.

25. Daniel Zwerdling, "Green Revolution Trapping India's Farmers in Debt," npr.org, April 14, 2009, www.npr.org/templates/story/story.php?storyId=102944731; "Neerjaal: India's First Holistic Water Resource Management System," Digital Empowerment Foundation, n.d., http://defindia.net/section_full_story.asp?id=427; "Concern over Depletion of Groundwater Sources," Hindu Business Line, September 19, 2007, www.thehindubusinessline.com/2007/09/19/stories/2007091952411000.htm.

26. "China's Water Shortage"; Richard Evans, "North China Groundwater Management Strategy," abstract, Sinclair Knight Merz, July 23, 2002, www.skmconsulting.com/Markets/environmental/North_China_Groundwater

_Management_Strategy.htm; Zijun Li, "Expert: Half of Chinese Cities Have Polluted Groundwater," Worldwatch Institute, December 6, 2005, www.worldwatch.org/node/1060; European Environment Agency, "Inland Waters," chap. 5 of *Europe's Environment—The Dobris Assessment,* State of the Environment Report 1, 1995, http://reports.eea.europa.eu/92–826–5409–5/en/page005new.html (to download full text, see link to zipped file at bottom of page).

27. International Assessment of Agricultural Knowledge, Science and Technology for Development, "Executive Summary of Synthesis Report of the International Assessment of Agricultural Knowledge, Science and Technology for Development (IAASTD)," 2008, p. 4, www.agassessment.org/docs/IAASTD_EXEC_SUMMARY_JAN_2008.pdf.

4. PEAK LAND

1. Ingrid Richardson, "Fertilisers, a Precious Commodity," Rabobank Global Focus Report, summer 2007, fig. 1.

2. Ghislain de Marsily, "Water, Climate Change, Food and Population Growth," *Revue des Sciences de l'Eau* 21, no. 2 (2008): 111–28.

3. The statistics I cite here and elsewhere in the chapter, unless attributed to a specific source, are numbers I generated using FAOSTAT, an online databank maintained by the Food and Agriculture Organization of the United Nations (FAO). Interested readers will find land area data at http://faostat.fao.org/site/377/default.aspx#ancor. To calculate worldwide farmland, for example, use World+, agricultural (all forms of agriculture, including livestock) or arable (crop production only) area, and the year. (The numbers generated will not be identical to mine because the FAO constantly updates these data.) For global population statistics, go to http://faostat.fao.org/site/550/default.aspx#ancor, and put in World+ and year (at far right). For food consumption, go to http://faostat.fao.org/site/609/DesktopDefault.aspx?PageID=609#ancor and put in World+, grand total, and kg/cal.day to get individual consumption. For the global consumption figures that I use, I multiplied population by caloric intake per person, that is, 6.5 million people × 2,800 kcal/person/day = 18.2 billion kcal/day.

4. Alex Evans, "The Feeding of Nine Billion: Global Food Security for the 21st Century, a Chatham House Report," January 26, 2009, www.humansecuritygateway.info/documents/CHATHAM_FeedingNineBillion_GlobalFoodSecurity21stCentury.pdf; Ama Biney, "Land Grabs—Another Scramble for Africa," AllAfrica.com, September 17, 2009, http://allafrica.com/stories/printable/200909180570.html.

5. International Soil Reference and Information Centre, *Global Assessment of Human-Induced Soil Degradation* (Wageningen, Netherlands: International Soil Reference and Information Centre, 1990), www-cger.nies.go.jp/grid-e/gridtxt/grid15.html. To look up the state of soil degradation in a particular country in 1990, go to www.fao.org/landandwater/agll/glasod/glasodmaps.jsp?country=%25&search=Display+map+%21.

6. Z. G. Bai, D. L. Dent, L. Olsson, and M. E. Schaepman, "Global Assessment of Land Degradation and Improvement. 1. Identification by Remote

Sensing," Report 2008/01, ISRIC–World Soil Information, Wageningen, Netherlands, 2008, p. i, www.isric.org/isric/webdocs/docs/Report%202008_01_GLADA%20international_REV_Aug%202008.pdf.

7. Ibid.

8. Ibid., p. 23.

9. Ibid., pp. 28, 29.

10. UN Environment Programme, "Don't Desert Drylands," message on World Environment Day, June 5, 2006, www.unep.org/wed/2006/downloads/PDF/UNEPWEDMessage06_eng.pdf.

According to Uriel Safriel and Zafar Adeel, "Drylands cover about 41% of Earth's land surface and are inhabited by more than 2 billion people (about one third of world population)." See their "Dryland Systems," chap. 22, p. 625, in *World Millennium Ecosystem Assessment, 2005,* vol. 1, *Current State and Trends Assessment,* www.maweb.org/documents/document.291.aspx.pdf. My extrapolation that one-third of the world's population will still live in deserts in 2050 is conservative.

11. Michael A. Stocking, "Land Degradation in the World's Most Acutely Affected Areas," lead presentation at UN Forum on Sustainable Land Management, New York, March 27, 2007, www.unuony.hypermart.net/seminars/2007/LandManagement/presentations/PaperStocking.pdf; S. M. Alam, R. Ansari, and M. Athar Khan, "Saline Agriculture and Pakistan," Industry and Economy, May 8–21, 2000, www.pakistaneconomist.com/issue2000/issue19&20/i&e3.htm.

12. H. R. von Uexküll and E. Muttert, "Global Extent, Development, and Economic Impact of Acid Soils," abstract of paper presented at the International Symposium on Plant-Soil Interactions at Low pH, September 12–16, 1993, http://cat.inist.fr/?aModele=afficheN&cpsidt=3543620.

13. Cooperative Research Centre for Contamination Assessment and Remediation of the Environment, "World Facing Arsenic Poisoning Calamity," February 2008, www.crccare.com/view/index.aspx?id=14938.

14. "Rising Ozone Pollution Threatens Crop Yields," *Nature,* May 5, 2005, excerpted at SciDev.Net, www.scidev.net/en/news/rising-ozone-pollution-threatens-crop-yields.html.

15. Xu Qi, "Facing Up to 'Invisible Pollution,'" chinadialogue.net, January 29, 2007, www.chinadialogue.net/article/show/single/en/724-Facing-up-to-invisible-pollution-.

16. Bruce Sundquist, "Urbanization-Caused Topsoil (Cropland) Loss," chap. 6 in *Topsoil Loss and Degradation,* http://home.windstream.net/bsundquist1/index.html.

17. Ibid.

18. "Japan: Black Day at Narita Airport," *Time,* April 10, 1978, www.time.com/time/magazine/article/0,9171,948067-2,00.html.

19. John Garnaut, "Farm Grab Sows Seeds of Rebellion," theage.com.au, January 2, 2008, www.theage.com.au/news/world/farm-grab-sows-seeds-of-rebellion/2008/01/01/1198949815542.html.

20. United Nations Environment Programme, "Plant for Planet: Questions and Answers," n.d., www.unep.org/billiontreecampaign/FactsFigures/QandA/index.asp.

21. Alan Harten, "Norway Offers Brazil $1 Billion to Save the Amazon," FairHome, September 17, 2008, www.fairhome.co.uk/2008/09/17/norway-offer-brazil-1-billion-to-save-the-amazon/.

22. Intergovernmental Panel on Climate Change, "Climate Change 2007: Synthesis Report," Summary for Policymakers, November 2007, www.ipcc.ch/pdf/assessment-report/ar4/syr/ar4_syr_spm.pdf; United Nations Environmental Programme, "Estimates of People Flooded in Coastal Areas in the 2080s as a Result of Sea-Level Rise and for Given Socio-Economic Scenarios and Protection Responses," June 2007, http://maps.grida.no/go/graphic/estimates-of-people-flooded-in-coastal-areas-in-the-2080s-as-a-result-of-sea-level-rise-and-for-given-socio-economic-scenarios-and-protection-responses.

23. Ross Garnaut, "Fateful Decisions," chap. 24 in *The Garnaut Climate Change Review* (Cambridge: Cambridge University Press, 2008), www.garnautreport.org.au/#FinalReport.

24. Kelly Young, "Greenland Ice Cap May Be Melting at Triple Speed," *NewScientist,* August 10, 2006, www.newscientist.com/article/dn9717Melt.

25. Anthony Fischer, "Brazilian Cerrado: Current Status and Prospects as a Food Bowl for the World," unpublished report, 2008.

26. European Bank for Reconstruction and Development, "EBRD, FAO and Russian Ministry of Agriculture to Join Forces to Boost Public-Private Cooperation," press release, October 17, 2008, www.ebrd.com/new/pressrel/2008/081017.htm.

27. I have calculated these estimates of growth in food output from global production statistics provided by the FAO's Statistics Division at http://faostat.fao.org.

28. Karen McGhee, "Restoring China's Grasslands," *Partners Magazine,* November 2007–February 2008, pp. 10–15, www.aciar.gov.au/publication/term/36.

29. George Monbiot, "Small Is Bountiful," Monbiot.com, June 10, 2008, www.monbiot.com/archives/2008/06/10/small-is-bountiful/.

30. For more information about permaculture, see "What Is Permaculture?" Permaculture Research Institute of Australia, September 13, 2005, http://permaculture.org.au/what-is-permaculture/.

5. NUTRIENTS—THE NEW OIL

1. Andrew Grice, "Britain Declares War on Food Waste," (London) *Independent,* July 7, 2008, www.independent.co.uk/environment/green-living/britain-declares-war-on-food-waste-861250.html.

2. "The £20bn Food Mountain: Britons Throw Away Half of the Food Produced Each Year," (London) *Independent,* March 2, 2008, www.independent.co.uk/life-style/food-and-drink/news/the-16320bn-food-mountain-britons-throw-away-half-of-the-food-produced-each-year-790318.html.

3. "U.S. Wastes Half Its Food," FOOD navigator-usa.com, November 26, 2004, www.foodnavigator-usa.com/Financial-Industry/US-wastes-half-its-food.

4. "Rot—Or Not?" *World Vegetable Center Newsletter,* May 23, 2008, www.avrdc.org/morenews/2008/May/foodwaste.pdf.

5. J. Lundqvist, C. de Fraiture, and D. Molden, "Saving Water: From Field to Fork—Curbing Losses and Wastage in the Food Chain," Stockholm International Water Institute Policy Brief, 2008, www.siwi.org/sa/node.asp?node=343.

6. Vaclav Smil, "Population Growth and Nitrogen: An Exploration of a Critical Existential Link," *Population and Development Review* 17 (1991): 569–601; Smil, "Phosphorus: Global Transfers," pp. 536–42 in Ted Munn, ed., *Encyclopedia of Global Environmental Change* (New York: Wiley, 2002).

7. M. Nelson and M. Mareida, "Environmental Impacts of the CGIAR: An Assessment," Doc. no. SDR/TAC:IAR/01/11, presented to the Mid-Term Meeting of the Consultative Group on International Agricultural Research, May 21–25, 2001, Durban, South Africa.

8. Ingrid Richardson, "Fertilisers, a Precious Commodity," Rabobank Global Focus Report, summer 2007, fig. 5.

9. O. C. Bockman, O. Kaarstad, O. H. Lie, and I. Richard, "Agriculture and Fertilisers: Fertilisers in Perspective," Norske Hydro, Oslo, 1990; Doris Blaesing, "Controlling Nitrogen Losses—Can We Learn from the European Experience?" *Proceedings of the Australian Agronomy Conference,* Geelong, February 2003, www.regional.org.au/au/asa/2003/c/8/blaesing.htm.

10. Smil, "Phosphorus: Global Transfers."

11. R. P. Narwal, B. R. Singh, and R. S. Antil, "Soil Degradation as a Threat to Food Security," in Rattan Lal, David O. Hansen, Norman Uphoff, and Steven Slack, eds., *Food Security and Environmental Quality in the Developing World* (New York: CRC Press, 2002), p. 95.

12. Johan Rockström et al., "Planetary Boundaries—A Safe Operating Space for Humanity," *Nature,* September 23, 2009, www.nature.com/nature/journal/v461/n7263/full/461472a.html; Z. X. Tan, R. Lal, and R. D. Weibe, "Global Soil Nutrient Depletion and Yield Reduction," *Journal of Sustainable Agriculture* 26, no. 1 (2005): 133, 136, 141.

13. James Murray, "IEA Official: Peak Oil Is Closer Than We Think," BusinessGreen.com, August 4, 2009, www.businessgreen.com/business-green/news/2247241/iea-official-peak-oil-closer.

14. Richardson, "Fertilisers, a Precious Commodity," p. 6.

15. Plant Production and Protection Division, Food and Agriculture Organization, "Plant Nutrients: What We Know, Guess and Do Not Know," March 27, 2007, www.fao.org/ag/AGL/agll/plnu_know.stm#_Toc30481155.

16. Alex Evans, "The Global Fertilizer Crisis," GlobalDashboard, August 8, 2008, www.globaldashboard.org/scarcity/the-global-fertiliser-crisis/.

17. "The World Food Situation: An Overview," report prepared for Annual General Meeting of the Consultative Group on International Agricultural Research, Marrakech, December 6, 2005, www.ifpri.org/pubs/agm05/jvbagm2005.asp.

18. See Smil, "Phosphorus: Global Transfers."

19. People today have forgotten the millions who died of hunger in the towns and cities of Europe and Asia during and after World War II, when food supplies failed or were taken by the armies.

20. See, for example, Richard Kula, "Green Roofs and Maximizing Credits under the LEED Green Building System," *Green Roof Infrastructure Monitor,* Spring 2005, excerpted at greenroofs.org, www.greenroofs.org/index.php?option=com_content&task=view&id=26&Itemid.=40.

21. See "Sky Water Solution," Rolex Awards for Enterprise, 2002, http://rolexawards.com/en/the-laureates/makotomurase-the-project.jsp; and National Water Commission, Government of Australia, "Managed Aquifer Recharge," www.nwc.gov.au/www/html/521-managed-aquifer-recharge.asp.

22. Christine Furedy, Virginia Maclaren, and Joseph Whitney, "Reuse of Waste for Food Production in Asian Cities: Health and Economic Perspectives," in Mustafa Koc, Rod MacRae, Luc J. A. Mougeot, and Jennifer Welsh, eds., *For Hunger-proof Cities: Sustainable Urban Food Systems* (Ottawa: IDRC Books, 2000), www.idrc.ca/en/ev-30609–201–1-DO_TOPIC.html.

23. N. E. Means, C. J. Starbuck, R. J. Kremer, and L. W. Jett, "Effects of a Food Waste-Based Soil Conditioner on Soil Properties and Plant Growth," *Compost Science and Utilization* 13 (Spring 2005): 116–21.

24. R. Naidu, personal communication, October 10, 2008.

25. Smil, "Phosphorus: Global Transfers."

26. See Australian Artificial Photosynthesis Network, "Artificial Photosynthesis: A National Research Priority," August 21, 2002, http://74.125.47.132/search?q=cache:Ras2DE6mqrsJ:www.dest.gov.au/sectors/research_sector/policies_issues_reviews/key_issues/national_research_priorities/documents/rtf49p_rtf.htm%3Fwbc_purpose%3Dbasic%2523%2523%2523%2523%2523+%22%E2%80%9Cartificial+photosynthesis%E2%80%9D+food&cd=4&hl=en&ct=clnk&gl=us.

6. TROUBLED WATERS

1. Stuart Richey, interview by author, April 18, 2007.

2. Julian Anthony Koslow, *The Silent Deep: The Discovery, Ecology and Conservation of the Deep Sea* (Sydney: University of New South Wales Press, 2007).

3. Ibid.

4. Richard Black, " 'Only Fifty Years Left' for Sea Fish," BBC News, November 2, 2006, http://news.bbc.co.uk/2/hi/science/nature/6108414.stm.

5. Finn Lynge, *Arctic Wars: Animal Rights and Endangered Peoples* (Hanover, N.H.: Dartmouth College Press, 1992).

6. Fisheries and Aquaculture Department, Food and Agriculture Organization of the United Nations, "State of the World's Fisheries," 2006, table 1, www.fao.org/docrep/009/A0699e/A0699E00.HTM; Deakin University, "Our Pets' Gourmet Tastes Are Putting Pressure on Dwindling Fish Stocks," press release, August 25, 2008, www.deakin.edu.au/news/upload/250808petfoodindustry.pdf.

7. Cornelia Dean, "Study Sees 'Global Collapse' of Fish Species," *New York Times,* November 3, 2006, www.nytimes.com/2006/11/03/science/03fish.html

?ex=1320210000&en=1cbe6153c8bdfebd&ei=5090. Worm and his colleagues reported their findings in "Impacts of Biodiversity Loss on Ocean Ecosystem Services," *Science,* November 3, 2006, pp. 787–90.

8. Dean, "Study Sees 'Global Collapse.' "

9. Steven Murawski, Richard Methot, and Galen Tromble, letter to the editor, *Science,* June 1, 2007, p. 1281.

10. International Bank for Reconstruction and Development, "The Sunken Billions: The Economic Justification for Fisheries Reform," World Bank, October 2008, http://siteresources.worldbank.org/EXTARD/Resources/336681-1224775570533/SunkenBillionsFinal.pdf; Erik Stokstad, "Privatization Prevents Collapse of Fish Stock, Global Analysis Shows," *Science,* September 19, 2008, p. 1619.

11. Thor Lassen, "Fisheries Bright Future," Ocean Trust Fish Facts, n.d., www.oceantrust.org/fishfacts.htm; "Scallop Comeback," *Fish* 15, no. 3 (September 2007), www.frdc.com.au/pub/news/153.php?article=1.

12. Fisheries and Aquaculture Department, "State of the World's Fisheries."

13. Ibid.

14. Meryl Williams, "Enmeshed: Australia and South East Asia's Fisheries," Lowy Institute, 2007, www.lowyinstitute.org/Publication.asp?pid=714.

15. Joachim von Braun et al., "Overview of the World Food Situation," brief prepared for the Annual General Meeting of the Consultative Group on International Agricultural Research, Nairobi, October 29, 2003, www.ifpri.org/pubs/speeches/20031029vonbraun.htm.

16. Based on data provided in Fisheries and Aquaculture Department, "State of the World's Fisheries."

17. Meryl Williams, personal communication, September 3, 2008.

18. Royal Society, "Ocean Acidification Due to Increasing Atmospheric Carbon Dioxide," Royal Society policy document 12/05, June 2005, p. vi, http://royalsociety.org/displaypagedoc.asp?id=13539.

19. Quoted in Julian Cribb, "Acid Oceans: The Silent Menace," *Fish* 16, no. 1 (March 2008), www.frdc.com.au/pub/news/161.php?article=28.

20. Quoted in ibid.

21. Quoted in ibid.

22. Ibid. Veron is the author of *A Reef in Time* (Boston: Belknap/Harvard, 2008).

23. Quoted in Cribb, "Acid Oceans."

24. Jeremy Jackson, "Ecological Extinction and Evolution in the Brave New Ocean," *Proceedings of the National Academy of Sciences* 105, supplement 1 (August 12, 2008): 11458–65, www.pnas.org/content/105/suppl.1/11458.full.

7. LOSING OUR BRAINS

1. "Threat Level Rising," *CIMMYT E-News* 3, no. 12 (December 2006), www.cimmyt.org/english/wps/news/2006/dec/wheatRust.htm; International Maize and Wheat Improvement Center (CIMMYT), "Dangerous Wheat Disease

Jumps Red Sea," press release, January 11, 2007, www.cimmyt.org/english/wps/nrs/GRI_RustSpread.pdf.

2. "The Wheat Rust Threat," *CIMMYT E-News* 3, no. 10 (October 2006), www.cimmyt.org/english/wps/news/2006/oct/thewheatrust.htm.

3. To learn more about these unsung heroes who keep the world fed, see Consultative Group on International Agricultural Research (CGIAR), "Scientists of the CGIAR," 2005, www.cgiar.org/pdf/pub_scientistsofthecgiar_2005.pdf.

4. "Norman Borlaug," *New World Encyclopedia,* April 3, 2008, www.newworldencyclopedia.org/entry/Norman_Borlaug?oldid=685200; Paul Ehrlich, *The Population Bomb* (New York: Ballantine Books, 1968), p. 1, recently cited in Paul Ehrlich, "The Return of the Population Bomb," *Environmental Health News,* July 14, 2009, http://www.environmentalhealthnews.org/ehs/editorial/the-return-of-the-population-bomb.

5. David A. Raitzer, executive summary of "Benefit-Cost Meta-Analysis of Investment in the International Agricultural Research Centres of the Consultative Group on International Agricultural Research," report prepared for Science Council Secretariat, Food and Agriculture Organization of the United Nations, 2003, p. xv.

6. Alex Evans, "The Feeding of Nine Billion: Global Food Security for the 21st Century," a Chatham House Report, Royal Institute for International Affairs, 2009, www.chathamhouse.org.uk/files/13179_r0109food.pdf.

7. For a discussion of the plight of soil science, for example, see Mary E. Collins, "Where Have All the Soils Students Gone?" *Journal of Natural Resources and Life Sciences Education* 37 (2008): 117–24.

8. Tom Lumpkin, address on global food price crisis, Canberra, Australia, September 2008.

9. CGIAR, "CGIAR Change Management: Working Group 4—Funding Mechanisms," report, September 15, 2008, www.cgiar.org/changemanagement/pdf/wg4_final_paper_sept15_2008.pdf.

10. Philip G. Pardey et al., "Science, Technology and Skills," October 2007, report commissioned by the CGIAR Science Council as a background paper for the 2008 World Development Report of the World Bank, www.apec.umn.edu/faculty/ppardey/publications.html.

11. Ibid., p. 82.

12. International Rice Research Institute, "Global Food Situation at a Crossroads," September 19, 2008, http://beta.irri.org/news/index.php/200809195067/press-releases/2008/Global-food-situation-at-a-crossroads.html#top.

13. For her prepared remarks see Katherine Sierra, "Climate Change, Sustainable Agriculture and the Research Road Ahead," keynote address, pp. 4–8, in *Agriculture in a Changing Climate,* proceedings of the ATSE Crawford Fund Fourteenth Annual Conference, Canberra, September 3, 2008, www.crawfordfund.org/events/pdfs/conference08proceedings.pdf.

14. FAO, Declaration of the World Summit on Food Security, Rome, November 16–18, 2009, p. 5, http://www.fao.org/fileadmin/templates/wsfs/Summit/Docs/Final_Declaration/WSFS09_Declaration.pdf.

15. Bill and Melinda Gates Foundation, Agricultural Development Overview, 2009, www.gatesfoundation.org/agriculturaldevelopment/Pages/overview.aspx.

16. CGIAR, "CGIAR Change Management," p. 8.

17. Norman Borlaug, "The Green Revolution Revisited and the Road Ahead," Nobel anniversary lecture, September 26, 2002, p. 18, http://nobelprize.org/nobel_prizes/peace/articles/borlaug/borlaug-lecture.pdf.

18. See, for example, the Biological Initiative for Open Source, www.bios.net/daisy/bios/home.html.

19. International Service for the Acquisition of Agri-biotech Applications, "Global Status of Commercialized Biotech/GM Crops: 2006," ISAAA Brief 35–2006: Executive Summary, www.isaaa.org/RESOURCES/PUBLICATIONS/BRIEFS/35/EXECUTIVESUMMARY/default.html; CropGen, "Making the Case for Crop Biotechnology," October 13, 2005, www.cropgen.org/article_44.html. For an opposing view, see Sam Burcher, "Global GM Crops Area Exaggerated," Institute of Science in Society, January 29, 2007, www.i-sis.org.uk/GlobalGMCropsAreaExaggerated.php.

20. See, for example, Hilary King, "Diabetes in Adults Is Now a Third World Problem," *Journal of Community Eye Health* 9, no. 20 (1996), www.cehjournal.org/0953–6833/09/jceh_09_20_051.html.

21. See Bruce Sundquist, "The Green Revolution, Fertilizers, and Pesticides," chap. 7 in *The Earth's Carrying Capacity*, June 2009, http://home.windstream.net/bsundquist1/index.html.

22. CIMMYT, "Wild Wheat Relatives Help Boost Genetic Diversity," August 2004, www.cimmyt.org/english/wps/news/wild_wht.htm; Lumpkin, address on global food price crisis.

23. Dilantha Gunawardana, "Supercharging the Rice Engine," *Rice Today* 7, no. 3 (July–September 2008): 20–21, www.irri.org/publications/today/pdfs/7–3/RT_7–3_complete.pdf.

24. Borlaug, "Green Revolution Revisited," p. 2.

25. Agriculture and Consumer Protection Department, Food and Agriculture Organization of the United Nations, "Spotlight 2002: New Animal Disease Threats," June 2002, www.fao.org/Ag/magazine/0206sp1.htm.

26. For examples of knowledge-transfer techniques, see World Vegetable Center, "Report of the 7th External Program and Management Review," March 3, 2008, www.avrdc.org/publications/EPMR/7th/Final_2007_EPMR_Report_6_May_08–2nd.pdf.

8. EATING OIL

1. See Rachel Oliver, "All about: Fossil Fuels," CNN.com/Asia, March 17, 2008, http://edition.cnn.com/2008/WORLD/asiapcf/03/16/eco.food.miles/.

2. Mario Giampetro and David Pimentel, "The Tightening Conflict: Population, Energy Use, and the Ecology of Agriculture," 1994, http://dieoff.org/page69.htm.

3. The term *peak oil* was coined by the geologist M. King Hubbert, who accurately predicted that U.S. oil production would peak in the 1970s, though few

believed him at the time. See M. King Hubbert, "Nuclear Energy and Fossil Fuels," *American Petroleum Institute Drilling and Production Practice Proceedings* (Spring 1956): 5–75.

4. Colin Campbell and Jerry [Jeremy] Gilbert, "The Age of Oil—Beginning of the End?" presentation by Gilbert for Southern California Energy Conference "Our Energy Future," Los Angeles, March 10, 2006, www.scag.ca.gov/rcp/pdf/summit/Gilbert.pdf.

5. James Murray, "IEA Official: Peak Oil Is Closer Than We Think," businessgreen.com, August 4, 2009, www.businessgreen.com/business-green/news/2247241/iea-official-peak-oil-closer.

6. Ibid.; Pedro Prieto to invitees to ASPO [Australian Association for the Study of Peak Oil and Gas] International Conference, Barcelona, October 20–21, 2008, ASPO Australia, n.d., www.aspo-australia.org.au/latest/garnaut-consider-peak-oil-&-aspo-7-barcelona.html.

7. These figures are based on fertilizer consumption data for 1990 and 2002, available from FAOSTAT, an online databank maintained by the Food and Agriculture Organization of the United Nations (FAO). See http://faostat.fao.org/site/422/DesktopDefault.aspx?PageID=422.

8. Richard Heinberg, "What Will We Eat as the Oil Runs Out?" Lady Eve Balfour Lecture, November 22, 2007, posted at Global Public Media, December 3, 2007, http://globalpublicmedia.com/richard_heinbergs_museletter_what_will_we_eat_as_the_oil_runs_out.

9. Henry Kindall and David Pimentel, "Constraints on the Expansion of Global Food Supply," *Ambio* 23 (1994): 198–205.

10. Kjell Aleklett, "Economy and Climate on the Path Down from the Peak of Oil and Gas," *ScienceAlert,* November 2, 2009, http://www.sciencealert.com.au/opinions/20090211-20143.html.

11. Mark W. Rosegrant, Siwa Msangi, Timothy Sulser, and Rowena Valmonte-Santos, "Biofuels and the Global Food Balance," in Peter Hazell and R. K. Pachauri, eds., *Bioenergy and Agriculture: Promises and Challenges* (Washington, D.C.: International Food Policy Research Institute, 2006); Donald Mitchell, "A Note on Rising Food Prices," abstract, World Bank Policy Research Working Paper Series, July 1, 2008, http://ssrn.com/abstract=1233058; Aditya Chakrabourty, "Secret Report: Biofuel Caused Food Crisis," guardian.co.uk, July 3, 2008, www.guardian.co.uk/environment/2008/jul/03/biofuels.renewableenergy.

12. World Bank, "Biofuels: The Promise and the Risks," Development Report 2008, http://siteresources.worldbank.org/INTWDR2008/Resources/2795087-1191440805557/4249101-1191956789635/Brief_BiofuelPrmsRisk_web.pdf.

13. FAO, "Biofuels: Prospects, Risks and Opportunities," pt. 1 of *The State of Food and Agriculture 2008* (Rome: FAO, 2008), www.fao.org/docrep/011/i0100e/i0100e00.htm.

14. Joachim von Braun, "The World Food Situation: New Driving Forces and Required Actions," International Food Policy Research Institute, December 2007, table 5, www.ifpri.org/pubs/fpr/pr18.pdf.

15. FAO, "Current World Fertilizer Trends and Outlook 2011/12," 2008, p. 6, ftp://ftp.fao.org/agl/agll/docs/cwfto11.pdf.

16. According to John Hampsarum, a cotton and grains farmer in New South Wales, it takes about 2.5 kilograms of oilseed grain to produce 1 liter of biodiesel (personal communication, September 26, 2008). To fill an SUV's 60-liter tank with biodiesel takes 150 kilograms of oilseed grain. Wayne Meyer, a professor of irrigation at Adelaide University, says that it takes 1,650–2,200 (say, 2,000) liters of water to grow 1 kilogram of oilseeds. Therefore, 150 kilograms × 2,000 liters = 300,000 liters (79,248 U.S. gallons), or 300 tonnes (330 U.S. tons) of water (Wayne S. Meyer, "Water for Food—The Continuing Debate," Land and Water, Commonwealth Scientific and Industrial Research Organisation, Government of Australia, n.d., www.clw.csiro.au/publications/water_for_food.pdf).

17. Ron Leng, "The Morality of Biofuels," editorial, *ScienceAlert,* May 7, 2007, www.sciencealert.com.au/opinions/20070805-15733.html; Piers Forster et al., "Changes in Atmospheric Constituents and in Radiative Forcing," chap. 2, table 2.14, in Intergovernmental Panel on Climate Change, *IPCC Fourth Assessment Report, Working Group 1 Report: The Physical Basis* (Paris: IPCC, 2007), http://www.ipcc.ch/pdf/assessment-report/ar4/wg1/ar4-wg1-chapter2.pdf. The information about carbon release is based on figures that appear in Joseph Fargione et al., "Land Clearing and the Biofuel Carbon Debt," *Science,* February 29, 2008, pp. 1235–38, www.sciencemag.org/cgi/content/abstract/319/5867/1235.

18. Hamparsum, personal communication, September 26, 2008.

19. For a good review of these technologies, see A. C. Warden and V. S. Haritos, "Future Biofuels for Australia," Rural Industries Research and Development Corporation, Government of Australia, 2008, https://rirdc.infoservices.com.au/downloads/08–117.pdf.

20. International Energy Agency, *Biofuels for Transport: An International Perspective* (Paris: IEA, 2004), fig. 6.1, www.iea.org/textbase/nppdf/free/2004/biofuels2004.pdf.

21. "Waste into Fuel," Rolex Awards, 2008, http://rolexawards.com/en/the-laureates/alexisbelonio-the-project.jsp.

22. David Strahan, "Biofuel without Tears—But How Much?" Global Public Media, February 21, 2008, http://globalpublicmedia.com/biofuel_without_tears.

23. Eviana Hartman, "A Promising Oil Alternative: Algae Energy," *Washington Post,* January 6, 2008, p. N06, www.washingtonpost.com/wp-dyn/content/article/2008/01/03/AR2008010303907.html; "Uses of Algae as Energy Source, Fertilizer, Food, and Pollution Control," Oilgae.com, 2006, www.oilgae.com/algae/use/use.html.

24. Hartman, "A Promising Oil Alternative."

25. I knew a farmer who ran his electric fences and radio from batteries made out of a plastic bucket of fresh pig manure with two electrodes in it: there is no end to the inventiveness of farmers in this field.

26. Jodi Ziesemer, "Energy Use in Organic Food Systems," Natural Resources Management and Environment Department, FAO, 2007, www.fao.org/docs/eims/upload/233069/energy-use-oa.pdf.

27. Raj Patel, "Commentary: Food from the Grassroots," *Yes!* Fall 2008, http://yesmagazine.org/article.asp?ID=2863.

28. International Bank for Reconstruction and Development and World Bank, "Overview: What Can Agriculture Do for Development?" p. 1, in *World Development Report 2008: Agriculture for Development* (Washington, D.C.: IBRD and World Bank, 2007), http://siteresources.worldbank.org/INTWDR2008/Resources/2795087-1192111580172/WDROver2008-ENG.pdf.

29. Henry Saragih, "Via Campesina Statement at the UN General Assembly on the Global Food Crisis and the Right to Food," La Via Campesina, April 6, 2009, www.viacampesina.org/main_en/index.php?option=com_content&task=view&id=698&Itemid=27.

30. "Declaration of the Via Campesina Second Youth Assembly," October 21, 2008, www.viacampesina.org/main_en/index.php?option=com_content&task=view&id=619&Itemid=1.

31. International Assessment of Agricultural Knowledge, Science and Technology for Development, "Executive Summary of the Synthesis Report of the International Assessment of Agricultural Knowledge, Science and Technology for Development (IAASTD)," 2008, p. 3, www.agassessment.org/docs/IAASTD_EXEC_SUMMARY_JAN_2008.pdf.

32. John Williams, personal communication, November 2008.

33. Heinberg, "What Will We Eat?"

34. This section is based on David J. C. MacKay, *Sustainable Energy—Without the Hot Air* (Cambridge: UIT Cambridge, 2009), www.inference.phy.cam.ac.uk/sustainable/book/tex/cft.pdf. See the discussion that begins on p. 24.

9. THE CLIMATE HAMMER

1. Global Crop Diversity Trust, "Arctic Seed Vault," 2006, www.croptrust.org/main/arctic.php?itemid=211.

2. Cary Fowler, "Conserving Crop Diversity: Navigating Politics and Climate Change to Create a Global System," p. 16, in proceedings of the ATSE Crawford Fund Fourteenth Annual Conference, Canberra, September 3, 2008, www.crawfordfund.org/events/pdfs/conference08proceedings.pdf; Crawford Fund, "Climate-Ready Crops Needed to Prepare Australian and Pacific Rim Agriculture for Climate Change, Says Global Seed Expert," press release, September 3, 2008, p. 2, www.crawfordfund.org/events/pdfs/fowlerpr.pdf.

3. Food and Agriculture Organization of the United Nations (FAO), "Climate Change and Food Security," November 2007, United Nations joint press kit for Climate Change Conference, Bali, December 3–14, 2007, p. 1, www.un.org/climatechange/pdfs/bali/fao-bali07-6.pdf; FAO, Declaration of the World Summit on Food Security, Rome, November 16–18, 2009, p. 2, http://www.fao.org/fileadmin/templates/wsfs/Summit/Docs/Final_Declaration/WSFS09_Declaration.pdf.

4. FAO, "Climate Change and Food Security."

5. FAO, introduction to "Climate Change and Food Security: A Framework Document," 2007, p. 1, ftp://ftp.fao.org/docrep/fao/010/k2595e/k2595e00.pdf.

6. "Rice-Wheat Systems and Climate Change," www.knowledgebank.irri.org/croppingSystem/factsheets/rice-wheat%20systems%20and%20climate%20change.pdf, adapted from P. R. Grace, M. C. Jain, and L. W. Harrington, "Environmental Concerns in Rice-Wheat Systems," pp. 99–111, in *Proceedings of the International Workshop on Development of Action Program for Farm-Level Impact in Rice-Wheat Systems of the Indo-Gangetic Plains,* September 25–27, 2000, Rice-Wheat Consortium Paper Series 14, Rice-Wheat Consortium for the Indo-Gangetic Plains, New Delhi; Pedro Sanchez, "The Climate Change–Soil Fertility–Food Security Nexus," summary note for the International Food Policy Research Institute conference Sustainable Food Security for All by 2020, Bonn, 2001, www.ifpri.org/2020conference/PDF/summary_sanchez.pdf.

7. Fran Kelly, "Producing Food in a Changing Climate," *Breakfast,* ABC [Australia] Radio National, September 4, 2008, www.abc.net.au/rn/breakfast/stories/2008/2354924.htm.

8. G. C. Nelson, M. W. Rosegrant, et al., "Climate Change: Impact on Agriculture and Costs of Adaptation," International Food Policy Research Institute, Washington D.C., September 30, 2009, www.ifpri.org/publication/climate-change-impact-agriculture-and-costs-adaptation.

9. Catherine Brahic, "Collapse of Civilisations Linked to Monsoon Changes," *NewScientist,* January 2007, http://environment.newscientist.com/article/dn10884; R. James Woolsey, "Catastrophic Climate Change over Next Hundred Years," p. 86, in Kurt M. Campbell et al., "The Age of Consequences: Foreign Policy and National Security Implications of Global Climate Change," Center for Strategic and International Studies and Center for New American Security, Washington, D.C., November 2007, www.csis.org/media/csis/pubs/071105_ageofconsequences.pdf.

10. Campbell et al., "Age of Consequences," p. 8.

11. Executive summary, p. 3, of Byron Bates, Zbigniew W. Kundzewicz, Shaohong Wu, and Jean Palutikof, eds., "Climate Change and Water," technical paper of the Intergovernmental Panel on Climate Change, IPCC Secretariat, Geneva, 2008, www.ipcc.ch/ipccreports/tp-climate-change-water.htm.

12. Executive summary, fig. 13, of Günther Fischer, Mahendra Shah, Harrij van Velthuizen, and Freddy O. Nachtergaele, "Global Agro-ecological Assessment for Agriculture in the 21st Century," Research Report RR-02–02, International Institute for Applied Systems Analysis, Laxenburg, Austria, 2001, www.iiasa.ac.at/Research/LUC/SAEZ/index.html.

13. Mats Eriksson and Xu Jianchu, "Climate Change Impact on the Himalayan Water Tower," Stockholm Water Front, July 2008, p. 11, www.siwi.org/sa/node.asp?node=205.

14. Meteorological Office, United Kingdom, "Soil Moisture Content: Annual Map," www.metoffice.gov.uk/climatechange/science/projections/soil_annual.html.

15. Hansen et al. predict "a nearly ice free planet." Total ice-cap melting involves a sea-level rise of approximately 60–80 meters, according to the U.S. Geological Survey. This will probably not occur for one to two centuries. Bamber

et al. think that the melting of the West Antarctic ice sheet and Greenland by late in this century will contribute a sea-level rise of 3.3 to 3.5 meters. As the time frame addressed in this book is the current century, I have decided to use the nearer estimate. See abstract of James Hansen et al., "Target Atmospheric CO_2: Where Should Humanity Aim?" *Open Atmospheric Science Journal* 2 (2008): 217, www.bentham.org/open/toascj/openaccess2.htm; U.S. Geological Survey, "Sea Level and Climate," USGS Fact Sheet 002–00, January 2000, http://pubs.usgs.gov/fs/fs2–00/; Jonathan L. Bamber, Riccardo E. M. Riva, Bert L. A. Vermeersen, and Anne M. LeBroq, "Reassessment of the Potential Sea-Level Rise from a Collapse of the West Antarctic Ice Sheet," *Science*, May 15, 2009, pp. 901–3, www.sciencemag.org/cgi/content/abstract/324/5929/901.

16. Hansen et al., "Target Atmospheric CO_2."

17. Campbell et al., "Age of Consequences," pp. 55, 71, 81.

18. See, for example, Alister Doyle, "Faster Ice Melt Seen as Spur to UN Climate Treaty," Reuters, April 28, 2009, http://uk.reuters.com/article/oilRpt/idUKLS83314220090428.

19. Graeme Pearman, telephone interview by author, October 2008.

20. Campbell et al., "Age of Consequences," pp. 8–9.

21. See, for example, A. Barrie Pittock, *Climate Change: The Science, Impacts and Solutions* (Victoria, Australia: CSIRO Publishing, 2009); Lester R. Brown, *Plan B: Rescuing a Planet under Stress and a Civilization in Trouble* (Washington, D.C.: Earth Policy Institute, 2003), www.earth-policy.org/Books/PlanB_contents.htm; and George Monbiot, *Heat: How to Stop the Planet from Burning* (New York: Allen Lane, 2006). For more radical solutions to climate change, see Gwynne Dyer, *Climate Wars* (New York: Random House, 2008).

22. FAO, "Climate Change Adaptation and Mitigation: Challenges and Opportunities for Food Security," p. 9, paper prepared for High-Level Conference on World Food Security, Rome, June 2008, ftp://ftp.fao.org/docrep/fao/meeting/013/k2545e.pdf.

23. Kaddambot Siddique, personal communication, October 2008.

24. See Charles C. Mann, "Our Good Earth," *National Geographic,* September 2008, pp. 101–6, http://ngm.nationalgeographic.com/print/2008/09/soil/mann-text.

25. FAO, "Harnessing Carbon Financing to Boost Sustainable Farming," press release, October 28, 2008, www.fao.org/newsroom/en/news/2008/1000947/index.html; Carbon Coalition against Global Warming, "The Soil Carbon Manifesto," 2006, www.carboncoalition.com.au/; Christine Jones, "Soil Carbon's Impact on Water Retention," Soil Carbon and Water by Christine Jones (blog), March 5, 2006, http://soilcarbonwater.blogspot.com/.

26. FAO, "Climate Change Adaptation and Mitigation."

27. Paul B. Thompson, *The Spirit of the Soil* (New York: Routledge, 1995), pp. 120–26; Holistic Management International, "History of Holistic Management?" n.d., www.holisticmanagement.org/n7/What_is/what_is_07.html.

28. David Kemp, personal communication, November 17, 2008.

29. The figure I am using, 43.3 million tonnes, refers to all food, including imports. See Tara Garnett, "Cooking Up a Storm: Food, Greenhouse Gas Emissions and Our Changing Climate," Food Climate Research Network, Centre for

Environmental Strategy, University of Surrey, 2008, p. 12, www.fcrn.org.uk/frcnPubs/publications/PDFs/CuaS_web.pdf.

30. Klass Jan Kramer, Henri C. Moll, Sanderine Nonhebel, and Harry C. Wilting, "Greenhouse Gas Emissions Related to Dutch Food Consumption," *Energy Policy* 27, no. 4 (April 1999): 203–16.

31. Garnett, "Cooking Up a Storm," summary report, p. 16, www.sehn.org/tccpdf/food%20climate%20change%20Summary_web.pdf.

10. ELEPHANTS IN THE KITCHEN

1. Attributed to Robert Repetto in Shridath Ramphal, *Our Country, Our Planet* (Washington, D.C.: Island Press, 1992).

2. Derek Tribe, *Feeding and Greening the World: The Role of International Agricultural Research* (Wallingford, U.K.: CAB International and Crawford Fund for International Agricultural Research, 1994), p. 25.

3. Jared Diamond, *Collapse: How Societies Choose to Fail or Succeed* (New York: Penguin, 2006), p. 498.

4. Bjorn Lomborg, *The Skeptical Environmentalist: Measuring the Real State of the World* (New York: Cambridge University Press, 2001).

5. Vaclav Smil asserts that peak population growth was passed in 1967. See Vaclav Smil, "Feeding the World: How Much More Rice Do We Need?" pp. 21–23 in K. Toriyama, K. L. Heong, and B. Hardy, eds., *Rice Is Life: Scientific Perspectives for the 21st Century*, proceedings of the World Rice Research Conference, Tsukuba, Japan, November 4–7, 2004 (Manila: International Rice Research Institute, 2005).

6. Population Division of the Department of Economic and Social Affairs of the United Nations Secretariat, "Estimated and Projected Total Fertility for the World," table II.1, in *World Population Prospects: The 2008 Revision, Highlights* (New York: United Nations, 2009), www.un.org/esa/population/publications/wpp2008/wpp2008_text_tables.pdf; Leta Hong Fincher, "Experts: Falling Birth Rates to Cause 'Demographic Time Bomb,' " Voice of America, March 3, 2005, www.voanews.com/english/archive/2005-03/2005-03-02-voa41.cfm?CFID=55175015&CFTOKEN=49423438.

7. These figures are based on data I found in the UN's searchable database, http://esa.un.org/unpp/index.asp?panel=1, part of the *World Population Prospects* study cited in note 6. However, the fertility and life expectancy data were not available as this book was being prepared for publication, although they may have been reposted by the time you read this.

8. Gwynne Dyer, *Climate Wars* (Melbourne: Scribe, 2008), p. 55. Note that world population passed 6.8 billion in early 2009.

9. Kirsten Garrett, "Background Briefing: Population Control," ABC Radio National [Australia], September 21, 2008.

10. Justin McCurry, "Japanese Minister Wants 'Birth-Giving Machines,' AKA Women, to Have More Babies," *Guardian,* January 29, 2007, www.guardian.co.uk/world/2007/jan/29/japan.justinmccurry; Michelle Tsai, "What

Can Government Do to Make Fertility Rates Go Up?" *Slate,* September 13, 2007, www.slate.com/id/2173901/.

11. See, for example, the abstract for Alberto Minujin and Enrique Delamonica, "Mind the Gap! Widening Child Mortality Disparities," *Journal of Human Development and Capabilities* 4, no. 3 (November 2003): 397–418, www.informaworld.com/smpp/content~content=a713678127~db=all~order=page; and Gary S. Becker, Kevin M. Murphy, and Robert Tamura, "Human Capital, Fertility and Economic Growth," *Journal of Political Economy* 98, no. 5 (October 1990): S12–37.

12. A point made by Robin Baker in *Sperm Wars: The Science of Sex* (New York: Basic, 1996).

13. See William Rees and Mathis Wackernagel, "Ecological Footprints and Appropriated Carrying Capacity: Measuring the Natural Capital Requirements of the Human Economy," in A-M. Jansson, M. Hammer, C. Folke, and R. Costanza, eds., *Investing in Natural Capital: The Ecological Economics Approach to Sustainability* (Washington, D.C.: Island Press, 1994); and Mathis Wackernagel and William Rees, *Our Ecological Footprint: Reducing Human Impact on the Earth* (Philadelphia: New Society, 1995).

14. "Ecological Footprint and Biocapacity, 2005," updated as of October 2008 with data from National Footprints Account 2008 edition. To access this table, go to Global Footprint Network, "World Footprint: Do We Fit on the Planet?" www.footprintnetwork.org/download.php?id=509, scroll to the bottom, and click on "2008 Data Tables."

15. "Special Announcement: Living Planet Report 2006 Outlines Scenarios for Humanity's Future," *Footprint Network News,* newsletter of the Global Footprint Network, n.d., www.footprintnetwork.org/gfn_sub.php?content=global_footprint.

16. Tim Flannery, "Now or Never: A Sustainable Future for Australia?" *Quarterly Essay*, no. 31 (September 2008): 2, www.quarterlyessay.com/pdf/qePDF/QE31_Chapter_One_Extract.pdf.

17. These 2003 figures are derived from FAOSTAT, the searchable database of the Food and Agriculture Organization of the United Nations (FAO). To use it, go to http://faostat.fao.org/site/610/DesktopDefault.aspx?PageID=610, and select the following elements: country, meat, food consumption (kg/cap/year), and the latest year for which world data are available (2003, in this case). The database will show the latest meat consumption figures for that country. Compare results with twenty-five years earlier.

18. See, for example, the case of China, as described in Mia MacDonald, "Opinion: Chinese Farms a Growing Challenge," Worldwatch Institute, October 20, 2008, www.worldwatch.org/node/5916?emc=el&m=160496&l=7&v=ea5c798974.

19. According to the FAO, "some 465 million tonnes must be produced annually by 2050, the great bulk of which in the developing countries"; see FAO, "World Agriculture: Towards 2030/2050," interim report of Global Perspective Studies Unit, Rome, June 2006, p. 51, www.fao.org/ES/esd/AT2050web.pdf. FAOSTAT puts 2007 world meat production at 285.7 million tonnes, which

represents an increase of 180 million tonnes between 2007 and 2050. The FAO's Animal Production and Health Division says, however, that "global production of meat is projected to more than double from 229 million tonnes in 1999/01 to 465 million tonnes in 2050"—or an increase of 222 million tonnes from 2000 to 2050. See FAO, executive summary of "Livestock's Long Shadow: Environmental Issues and Options," 2006, p. xx, ftp://ftp.fao.org/docrep/fao/010/a0701e/a0701e00.pdf.

20. FAO, *The State of Food and Agriculture 2007* (Rome: FAO, 2007), p. 123, ftp://ftp.fao.org/docrep/fao/010/a1200e/a1200e00.pdf.

21. Macdonald, "Chinese Farms a Growing Challenge."

22. Robin de Milano, "India 2020: Rise of the Elephant," Rabobank Special Scenario Studies, Utrecht, Netherlands, January 2007, p. 3.

23. Vaclav Smil, "Water News: Bad, Good and Virtual," *American Scientist* 96, no. 5 (September–October 2008): 406, 407.

24. Executive summary of Pew Commission on Industrial Farm Animal Production, "Putting Meat on the Table: Industrial Farm Animal Production in America," p. 17, Johns Hopkins University, Bloomberg School of Public Health, Baltimore, April 2008, www.ncifap.org/.

11. A FAIR DEAL FOR FARMERS

1. FAO, Declaration of the World Summit on Food Security, Rome, November 16–18, 2009, p. 1, http://www.fao.org/fileadmin/templates/wsfs/Summit/Docs/Final_Declaration/WSFS09_Declaration.pdf.

2. Alex Evans, "The Feeding of Nine Billion: Global Food Security for the 21st Century, a Chatham House Report," January 26, 2009, p. 6, www.humansecuritygateway.info/documents/CHATHAM_FeedingNineBillion_GlobalFoodSecurity21stCentury.pdf.

3. Klaus von Grebmer et al., "Global Hunger Index: The Challenge of Hunger 2008," International Food Policy Research Institute, October 2008, p. 23, www.ifpri.org/sites/default/files/publications/ghi08.pdf.

4. FAO, "Diouf Appeals for New World Agricultural Order," press release, November 19, 2008, www.fao.org/news/story/en/item/8569/icode/; "Farmers 'Ignored' at Food Summit," *Sydney Morning Herald,* June 7, 2008, http://news.smh.com.au/world/farmers-ignored-at-un-food-summit-20080607–2n3b.html; International Food Policy Research Institute, "High Food Prices: The What, Who, and How of Proposed Policy Actions," policy brief, May 2008, www.ifpri.org/pubs/ib/FoodPricesPolicyAction.pdf; Welthungerhilfe quotation in Klaus von Grebmer et al., "The Challenge of Hunger: The 2008 Global Hunger Index," IFPRI Issue Brief 54 , October 2008, p. 6, www.ifpri.org/pubs/ib/ib54.pdf.

5. Max Borders and H. Sterling Burnett, "Farm Subsidies: Devastating the World's Poor and the Environment," National Center for Policy Analysis, March 23, 2006, www.ncpa.org/pub/ba/ba547/; "IMF Presses World to Scrap Farm Subsidies," Agence France-Presse, September 19, 2002, www.globalpolicy.org/component/content/article/209/43024.html.

6. See Andy Stoeckel, "High Food Prices: Causes, Implications and Solutions," Rural Industries Research and Development Corporation, Government of Australia, publication no. 08/100, June 2008, https://rirdc.infoservices.com.au/downloads/08–100.pdf.

7. Author's neologism to describe the combined natural resources of the planet used for human food production.

8. FAO, "High-Level Conference on World Food Security: The Challenges of Climate Change and Bioenergy," n.d., www.fao.org/foodclimate/hlc-home/en/.

9. Group of Twenty, "Declaration of the Summit on Financial Markets and the World Economy," November 15, 2008, Washington, D.C., www.g8.utoronto.ca/g20/2008-leaders-declaration-081115.html.

10. Andy Stoeckel and Hayden Fisher, "Policy Transparency: Why Does It Work? Who Does It Best?" Rural Industries Research and Development Corporation, Government of Australia, July 2008, p. x, https://rirdc.infoservices.com.au/downloads/08–035.pdf.

11. Andy Stoeckel, interview by author, November 18, 2008.

12. Joachim von Braun, "The World Food Situation: An Overview," remarks prepared for Annual General Meeting of the Consultative Group on International Agricultural Research, Marrakech, Morocco, December 6, 2005, www.ifpri.org/pubs/agm05/jvbagm2005.asp.

13. George Monbiot, "Small Is Bountiful," Monbiot.com, June 10, 2008, www.monbiot.com/archives/2008/06/10/small-is-bountiful/.

14. See, for example, Monsanto, "Growing Hope in Africa," n.d., www.monsanto.com/responsibility/our_pledge/stronger_society/growing_hope_africa.asp.

15. Joachim von Braun, "Best Bets for Reducing Poverty and Hunger: Opportunities for Investment in Agricultural Research," International Food Policy Research Institute, press statement, October 10, 2008, www.ifpri.org/pressrel/2008/20081010.asp.

16. FAO, "Investment," How to Feed the World 2050, High-Level Expert Forum, Rome, October 12–13, 2009, www.fao.org/fileadmin/templates/wsfs/docs/Issues_papers/HLEF2050_Investment.pdf.

17. Guy Healy, "Barrier Reef Lands in Google Net," *Australian,* February 11, 2009, www.theaustralian.news.com.au/story/0,,25036440–12332,00.html?from=public_rss.

18. For U.S. death rates, see Ben Best, "Causes of Death," n.d., www.benbest.com/lifeext/causes.html#data_usa.

19. "Poor and Hungry Cannot Afford to Wait, World Bank President Says," World Bank press release, April 29, 2008, http://web.worldbank.org/WBSITE/EXTERNAL/NEWS/0,,contentMDK:21749013~pagePK:34370~piPK:34424~theSitePK:4607,00.html.

20. R. Engelman, at www.worldwatch.org/node/6257?emc=el&m=297190&l=4&v=ea5c798974.

21. See, for example, Evans, "Feeding of Nine Billion."

22. Ibid., p. 7.

Table 13 NUTRIENT YIELD RELATIVE TO LAND AND WATER REQUIREMENTS

	Energy (kcal)			Protein (g)			Vitamin A (µg RAE1)			Iron (mg)		
Food group	*Per 100g*	*Per ha of land*	*Per M3 of water*	*Per 100g*	*Per ha of land*	*Per M3 of water*	*Per 100g*	*Per ha of land*	*Per M3 of water*	*Per 100g*	*Per ha of land*	*Per M3 of water*
Cereals												
Wheat	339	9,464,202	1,898	13.7	382,477	77	0	0	0	3.88	108,322	22
Maize	365	18,143,420	2,701	9.4	467,255	70	0	0	0	2.71	134,709	20
Sorghum	339	4,998,894	N/A	11.3	166,630	N/A	0	0	N/A	4.4	64,882	N/A
Average	**348**	**10,868,839**	**2,300**	**11**	**338,787**	**73**	**0**	**0**	**0**	**4**	**102,638**	**21**
Legumes												
Beans	343	14,044,478	N/A	23.85	976,562	N/A	2	81,892	N/A	9.95	407,413	N/A
Vegetables												
Tomatoes	18	4,911,948	1,062	0.9	245,597	53	42	11,461,212	2,478	0.27	73,679	16
Cabbages	25	5,615,775	2,830	1.3	292,020	147	5	1,123,155	566	0.47	105,577	53
Chilies and peppers	40	6,122,240	N/A	2	306,112	N/A	59	9,030,304	N/A	1.2	183,667	N/A
Carrots	41	9,141,811	4,030	0.9	200,674	88	835	186,180,785	82,081	0.3	66,891	29
Others (e.g. spinach)	23	3,615,347	771	2.9	455,848	97	469	73,721,641	15,712	2.71	425,982	91
Average	**29**	**5,881,424**	**2,173**	**2**	**300,050**	**96**	**282**	**56,303,419**	**25,209**	**1**	**171,159**	**47**
Meat												
Bovine meat	276	594,504	166	15	32,310	9	0	0	0	5.67	12,213	3
Chicken meat	234	1,355,094	679	18.8	108,871	55	37	214,267	107	1.09	6,312	3
Average	**255**	**974,799**	**422**	**17**	**70,590**	**32**	**19**	**107,134**	**54**	**3**	**9,263**	**3**

SOURCES: Jackie Hughes et al., "Vegetables for More Food Using Less Resources," World Vegetable Center, September 24, 2008, based on FAOSTAT 2007 (yields of vegetables and cereals/grains); Ephraim Leibtag, "Corn Prices Near Record High, but What about Food Costs?" *Amber Waves,* February 2008, www.ers.usda.gov/AmberWaves/February08/Features/CornPrices.htm (yield of meat); C. de Fraiture et al., "Does International Cereal Trade Save Water? The Impact of Virtual Water Trade on Global Water Use," Comprehensive Assessment Research Report 4, Colombo, Sri Lanka, 2004 (grain yield); B. Decrausaz, "Virtual Water and Agriculture in the Context of Sustainable Development," Organization for Economic Cooperation and Development Workshop on Agriculture and Water, Adelaide, November 16, 2005 (meat yield).

12. FOOD IN THE FUTURE

1. Ajay Vashee, keynote address to Australian Council of Agricultural Journalists, Canberra, November 11, 2008, www.ifap.org/en/newsroom/documents/AjayVashee_NationalPressClubAustralia.pdf.

2. Jackie Hughes et al., "Vegetables for More Food Using Less Resources," World Vegetable Center, September 24, 2008, p. 2, in the author's personal files.

3. Ibid., p. 2.

4. Ibid.

5. Ibid., p. 3. Table 13 compares resource requirements and nutrient yields for some common foods.

6. For a good article on the need for a new global diet, see Moises Velasquez-Manoff, "Diet for a More-Crowded Planet: Plants," *Christian Science Monitor,* July 18, 2008, http://features.csmonitor.com/environment/2008/07/18/diet-for-a-more-crowded-planet-plants/.

7. J. Lundqvist, C. de Fraiture, and D. Molden, "Saving Water: From Field to Fork—Curbing Losses and Wastage in the Food Chain," Stockholm International Water Institute Policy Brief, 2008, p. 30, www.siwi.org/documents/Resources/Policy_Briefs/PB_From_Filed_to_Fork_2008.pdf.

8. Norman E. Borlaug, "Protection of Biodiversity Key to Boosting Food Production," Global Crop Diversity Trust, 2006, www.croptrust.org/main/whatpeople.php.

9. The Food and Agriculture Organization of the United Nations (FAO) has asked the world's governments to raise $30 billion for "rural infrastructure and increased agricultural productivity in the developing world." Mine is a global estimate, for all countries, and factors in the cost of sharing the knowledge across 1.8 billion farmers, an area the FAO tends not to emphasize sufficiently.

10. World Bank, "Emerging National Agendas for Agriculture's Three Worlds," chap. 10 in *World Development Report 2008: Agriculture for Development,* http://econ.worldbank.org/WBSITE/EXTERNAL/EXTDEC/EXTRESEARCH/EXTWDRS/EXTWDR2008/0,,contentMDK:21410054~menuPK:3149676~pagePK:64167689~piPK:64167673~theSitePK:2795143,00.html.

11. Muhammad Yunus, *Creating a World without Poverty: Social Business and the Future of Capitalism* (New York: PublicAffairs, 2008), p. 56.

12. Jacques Diouf, the director general of the FAO, urges "a new system of governance of world food security and an agricultural trade that offers farmers, in developed and developing countries alike, the means of earning a decent living." See FAO, "Diouf Appeals for New World Agricultural Order," press release, November 19, 2008, www.fao.org/news/story/en/item/8569/icode/.

Index

Text: 10/13 Sabon
Display: Franklin Gothic
Compositor: Westchester Book Group
Indexer: Kevin Milham
Cartographer: Bill Nelson